MW00529586

ANCIENT TIBET

ANCIENT TIBET

Research Materials

from

The Yeshe De Project

Dharma  Publishing

TIBETAN HISTORY SERIES

Ancient Tibet: Research Materials

Library of Congress Cataloging in Publication Data

Ancient Tibet.
(Tibetan history series; v.1) Includes Index.
1. Tibet (China)–History. I. Dharma Publishing
II. Series.
DS786.A68 1986 951'.5 86-24124
ISBN 0-89800-146-3

Typeset in Mergenthaler Trump
Printed and bound in the United States of America
by Dharma Press, California

9 8 7 6 5 4 3 2

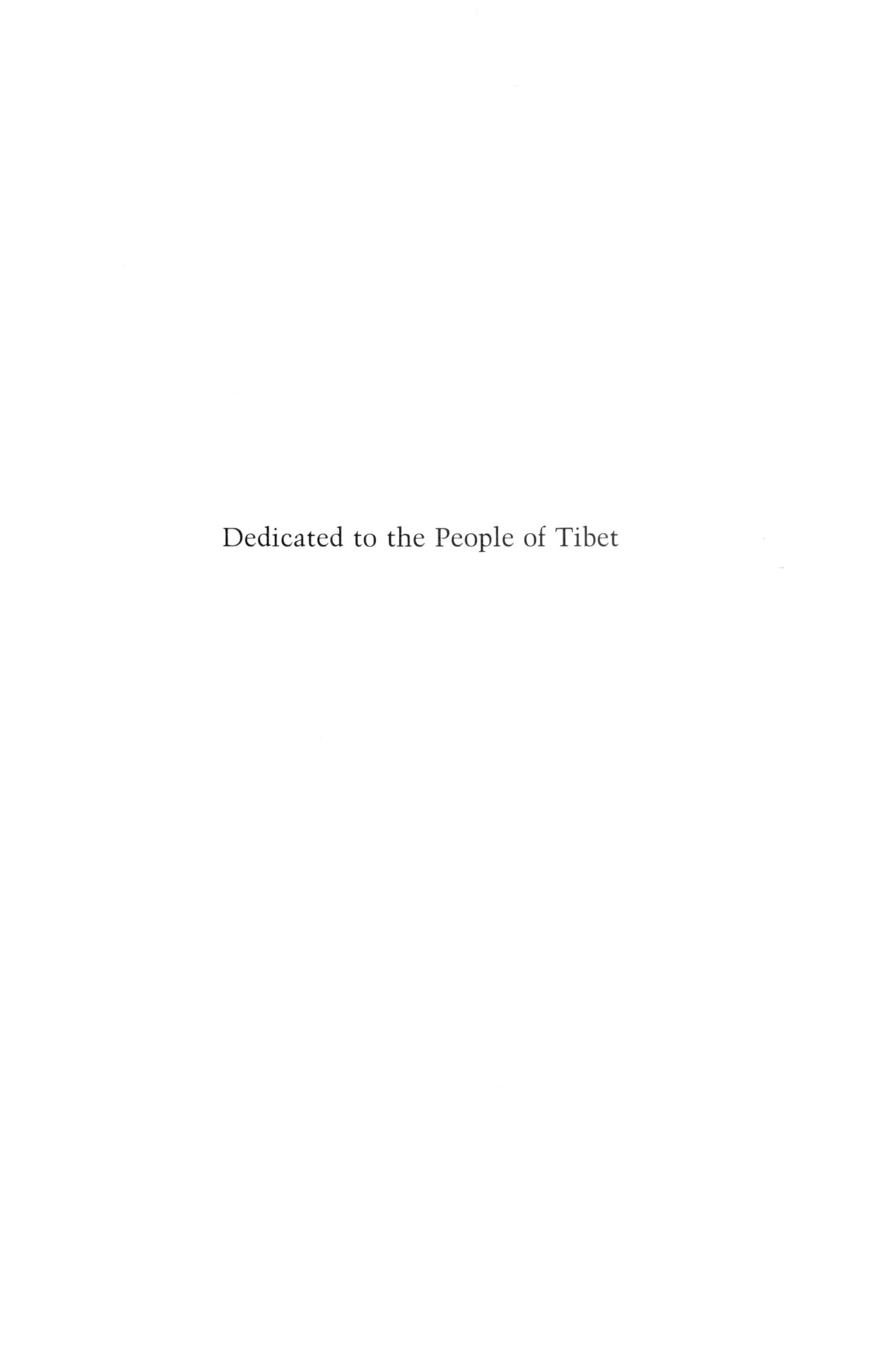

Dedicated to the People of Tibet

CONTENTS

Part One

Part Two

Part Three

PUBLISHER'S PREFACE

The history of Tibetan culture stretches back several thousand years to a time before the rise of modern civilizations. Surrounded by the immense power and beauty of the most dramatic landscape in the world, the Tibetan people created an independent and vigorous culture deeply in tune with the forces of nature. Located in the heart of Asia, Tibet very early became a meeting ground for Indian, Chinese, Central Asian, and even Western influences. The rich and complex civilization that resulted from these cultural interactions preserved and transmitted some of mankind's finest achievements.

Tibet's contributions to world civilization, however, are not well known. Westerners have often regarded Tibet as a kind of Shangri-la, an unknowable, timeless land hidden away beyond the mountain peaks. Few history books contain so much as a paragraph on Tibet, and many of Tibet's greatest kings and statesmen are known only to scholars in the field.

The history of these kings and the ancient Tibetan tribes is also little known to most Tibetans. For example, as a young man in Tibet, I received an excellent education, but history was not a part of my studies. As I grew older, I wished to know more about the ancient history of my fatherland. But I found few works that presented the important events of ancient times in any detail.

Those that did presupposed a great deal of familiarity with Tibetan history and historiography; lengthy debates would hide the main flow of events, with scholarly interpretations often interspersed among the facts. A simple but detailed review of ancient Tibetan history was hard to find.

I was also interested in how information in traditional Tibetan histories fit with old Tibetan records discovered early this century in Central Asia at Tun-huang. Histories of Tibet's ancient neighbors, which contained valuable references to events in Tibet, are also available now, for some of these records have been translated in the last century. But even newer books on Tibet contained only short summaries of this material. The actual texts of these records are in scholarly publications unavailable to general readers.

The idea of Dharma Publishing creating an introductory book on ancient Tibetan history began over six years ago in the course of our research for the Nyingma Edition of the bKa'-'gyur and bsTan-'gyur. It seemed that such a volume might be welcomed both by Tibetans interested in the origins of Tibetan civilization and by my Western students and friends.

To research the important events in ancient Tibetan history, the Dharma Publishing staff began to collect information from a variety of historical sources such as the Tun-huang annals, the T'ang annals, the Li-yul texts, and Arab histories. Over several years, different research teams consulted the Western translations and discussions of these sources. Friends of Dharma Publishing also brought interesting materials on Tibetan history to our attention.

Broadening our investigation into Tibetan history, we found that history led naturally to archeology, and archeology to geology; research expanded to include the history of the Tibetan plateau, the development of the Tibetan environment, and the earliest inhabitants of Tibet. When we realized how far back Tibet's history could be traced, and how intriguing this scientific information is in the light of traditional Tibetan accounts, we decided to include some of this material. We summarized the current research to the best of our ability and included simple diagrams so that the general reader could envision the processes that formed the land of Tibet. To integrate Tibetan history with the flow of world events, we included timelines as well as sketch maps, which are not precisely to scale, but illustrate the text. We hope these background materials will introduce Western perspectives to Tibetan readers and provide a broader setting in which to view Tibetan history.

As we were expanding the scope of our research, I continued to gather information from available authoritative Tibetan histories to compare their presentations. Working closely with me, Leslie Bradburn checked and organized the research materials and edited them into a coherent account.

In the course of compiling this book, we found that there are a number of unresolved questions in ancient Tibetan history. As with most civilizations, the ancient times are the least well documented. The further back in time we reach, the less we are able to find. Gaps in the records and conflicting information leave room for conjecture and different interpretations. Experts, both Tibetan and Western, will take different positions, making it difficult for nonscholars to know which is the best interpretation. Resolving such controversies or documenting one position versus another is a task better left to professional historians and scholars. We simply assembled the basic information from the most well-known sources, while pointing out interesting issues and offering an occasional opinion. Material from less well-known sources, both Western and Tibetan, that provides detailed information on certain points was not considered appropriate for an introductory volume. After further research, it may be possible to publish some of these materials.

This book is intended as a basic research tool for nonspecialists, and it should be considered neither authoritative nor complete. But we hope it will provide a survey of general information to give interested readers a good overview of ancient Tibetan history. A list of some sources for further study has been collected in the Appendix. In succeeding volumes, we hope to discuss more fully the development of Tibetan culture, a subject that this book only begins to explore. Topics for research include the transmission of the sixteen major and minor Tibetan Buddhist schools, the evolution of Tibetan art, literature, and medicine, and further historical and geographical studies.

We began work on the Tibetan history series after the publication of the Nyingma Edition of the bKa'-'gyur and bsTan-'gyur, expecting to complete the first volume in 1984. As this book rapidly expanded beyond our original intention, other pressing concerns began to interrupt our efforts. An early draft was set aside to complete several translations — The Voice of the Buddha, Mother of Knowledge, and The Marvelous Companion. When we began again, we decided to check our original research against additional Tibetan sources. But my time was soon taken by work on a comprehensive volume on the Odiyan center, Copper Mountain Mandala. By the summer of 1985, the book was almost ready. But preparation for publication was postponed again when an extensive art project for Odiyan required all our attention. By the time that project was complete, I had left on an extended visit to my family and monastery in Tibet. It now seems best not to delay publication any longer, even though I have not been able to give this volume as much attention as I would have liked. Additional Tibetan sources could be compared and many more accounts cross-checked, but such a project will have to wait for a later volume.

Although it is just an introduction, Ancient Tibet is a gesture of appreciation to the land where I was born and to its people. In the last few decades, Tibet has undergone dramatic, even drastic changes. My hope is that this volume will encourage the understanding of the value of the traditional culture while it still remains, and remind Tibetans of their ancient heritage.

Tarthang Tulku

PART ONE

THE LAND

PERSPECTIVE

THE SCIENTIFIC POINT OF VIEW

The Tibetan plateau is the highest land on the face of the earth, a dry steppe surrounded by snow-capped mountain peaks towering thousands of feet above the rest of Asia. Tibetan histories written many centuries ago describe a great transformation that changed the face of Tibet as the mountains rose and the land emerged from the sea. Now modern earth scientists agree that Tibet has not always been the Roof of the World. Recent research suggests that millions of years ago, long before the beginning of recorded human history, the plateau was forested lowland lying on the shore of an ancient tropical sea.

With the development of modern geology, the formation of the plateau has become the subject of scientific study. Today scientists are attempting to understand the ancient past of Tibet in relation to the geologic history of the whole earth. The first section of this book reports their findings and theories.

The modern scientific perspective emerged quite recently from older traditions and soon became almost unquestioned in the West. Modern science began in the seventeenth century when René Descartes and Francis Bacon laid the foundation for experimental science. New fields of inquiry and study arose: chemistry from alchemy, biology and physics from natural philosophy, astronomy from astrology. By the eighteenth century, geology was beginning to take shape as a specialized science of earth history. Today, the study of the origin, structure, and evolution of the earth has become a huge branch of science, combining the efforts of astronomers, physicists, mathematicians, earth scientists, and life scientists.

In the twentieth century, a new understanding of the origin of the world, indeed of the whole universe, began to take shape. The General Theory of Relativity, originally published seventy years ago by Albert Einstein, inspired new ideas about the age and origin of the universe.

When seventeenth century Christian theologians first calculated the age of the earth, they concluded that the earth had been created less than six thousand years ago. By the eighteenth century, scientists were suggesting an age of 75,000 years. Modern cosmologists now believe that the earth took shape over 4000 million years ago, while the date for the beginning of the universe has been estimated to be 14,000 to 15,000 million years ago.

This newest hypothesis of Western cosmologists is based on evidence that the universe is expanding. In 1929 it was determined that galaxies are moving away from each other. This suggested that in the earliest stages of the universe, space and matter had been condensed almost infinitely and had expanded explosively. In 1965 scientists detected a background radiation throughout space, which they took to be the echo of that original event.

The nature of this event, the origin of matter, space, energy, and even time itself, has baffled and intrigued cosmologists, who have joined together with quantum physicists to develop a scientific understanding of the most ancient and basic questions: What does it mean to speak of the time when time began or the point where space originated?

Although these underlying questions have in no way been resolved, a "standard model" of the earliest moments of the universe has become widely accepted by cosmologists. This scenario goes as follows: As the universe expanded at extremely high temperatures, elementary particles of matter arose from pure energy. Almost simultaneously, their combinations created protons and neutrons, the building blocks of atoms. Within another fraction of a second, the universe had expanded to the size of our solar system.

When the universe was still only minutes old, helium nuclei were created as pairs of protons and neutrons joined together. Several hundred thousand years passed before the universe had cooled to temperatures of a few thousand degrees, allowing electrons to bond to protons and neutrons to form the simplest atoms of hydrogen and helium. Space began to clear, and the universe blazed with light.

At this point in the story, the cosmologists and physicists step aside, and the astronomers tell the history of the stars and the solar system. About 1000 million years after the initial explosion, spinning clouds of hydrogen and helium gas drew together due to gravitational attraction, and huge clusters of galaxies began to form. Within the galaxies, smaller clouds collapsed, their temperature increasing tremendously, until they began to shine as new stars. Burning first hydrogen and helium, these star "furnaces" eventually produced carbon, oxygen, and other heavier elements.

As the ancient stars aged, some became exploding supernovas that scattered their material into space where it joined other clouds of matter to form new stars. Astronomers refer to our sun as a second or third generation star: The elements composing the sun and its planets are the "ashes" of stars that exploded in distant parts of the universe in ages past.

The star we call our sun began forming over four and a half billion (4500 million) years ago. As clouds of dust and gas accumulated around this new star, primitive planets took shape. The growing earth began to heat up, so that a partly solid, partly molten body formed. Over many millions of years, heavier material such as iron and nickel sank down toward the center of the new earth while lighter, granitelike rock made its way toward the surface. As the planet aged, its outer layers cooled and solidified into a thin crust. But deeper layers have remained hot and deformable, and the core of the earth is still partly molten even today.

The young earth lacked an atmosphere that would protect it against stray cosmic debris, and thus its surface was often bombarded with comets and asteroids, which delivered additional chemical elements to the new planet. In this era the cratered and barren earth must have resembled the moon. Over time, a primitive atmosphere formed as gases such as water vapor and carbon dioxide streamed forth from the new planet. These gases were held close to the surface by the earth's gravitational field, forming the first clouds. When the crust cooled, rain began to pour down upon the newly formed granite mountains, producing soil and sediment, and filling up depressions in the crust to make seas.

Science has only recently accounted for the processes that formed the present surface of the planet. As early as the seventeenth century Francis Bacon had noted that the coastlines on either side of the Atlantic Ocean could be fitted together like pieces of a puzzle. In the beginning of the twentieth century, Alfred Wegener proposed that the continents had drifted to their present positions over long periods of time. But no one could imagine how the continents had moved. By 1970, however, earth scientists had worked out a coherent theory, known as plate tectonics, to explain the building of mountains, the formation of oceans, and the shapes and locations of the modern continents.

The geologists bring us up in time to the formation of the land of Tibet. According to plate tectonics, the outer shell of the earth is divided into a number of separate moving plates that float upon the deeper, deformable layers of the earth, carrying both the continents and the ocean floor. The plates are slowly but constantly in motion, sliding past each other at the rate of two to eight centimeters per year. These motions account for the major features of the earth's surface.

As plates move apart from each other, molten lava from the deeper layers of the earth rises up through the rift between the plates. When lava hardens, it forms a new ocean basin between the separating plates, and thus an ocean "opens." The Atlantic

Ocean opened some 100 million years ago and is still growing wider as the sea floor continues to spread.

When a plate carrying a continent collides with a plate carrying ocean crust, the denser, thinner ocean plate bends down beneath the lighter, thicker continental plate and is partially melted in the deep layers of the earth. Thus oceans "close" as the two plates approach each other. The Pacific Ocean is gradually closing at both margins, and eventually Asia and America may collide with each other.

During such collisions, molten rock from remelted ocean crust rises up to the surface, creating volcanoes along the edge of the continent. The Andes mountains were built up from such eruptions as a plate carrying part of the Pacific Ocean descended beneath the South American plate.

No ocean crust is older than 200 million years because it is continually being destroyed and recycled. Rock as old as 4000 million years, however, has been found in continental crust, which resists this recycling. But this rock undergoes a different kind of transformation. Most of the continental crust on the planet has been involved in mountain-building events at one time or another over hundreds of millions of years of earth history.

Mountains are built up when plates carrying continents meet, for neither plate will bend beneath the other. In such a massive collision, the crust of each plate folds and crumples into ridges. The Himalayan ranges were thrust up during such a collision between Asia and India.

High ground built up by these plate movements is continually eroded and weathered, producing sediments that accumulate on lower ground and sink into the earth. Deep in the earth the rock melts and then may be thrown up to the surface again by plate movements. This creation and destruction continues to shape and reshape the face of the earth.

To track the movements of the continents through time, geologists have compared fossil and geologic records from sites around the world. The magnetism of the ocean floor and some continental rock can be used to determine ancient positions of land and sea, while the ages of certain types of rock can be calculated from rates of naturally occurring radioactive decay.

For example, the magnetism of the Indian Ocean floor indicates that hundreds of millions of years ago the Indian plate was located much farther south. It gradually drifted northward until it collided with the rest of Asia some 50 million years ago, wrinkling the shore of Asia into the Himalayas. This tremendous collision is the most recent mountain-building event in the history of the earth.

Reconstructing the past from fragmentary clues, the new theories of earth science let us envision the formation of the Tibetan plateau as an integral chapter in the ancient past of the planet. Tibet's mountain ranges and high plateau have a lifespan of millions of years. Throughout these long ages of time, the mountains are slowly eroding even as they rise, and the plateau is gradually shifting into a new shape. Even the substances composing the land—the clay and rock in the mountains, the water that fills the lakes—have endured cycles of change that trace back billions of years to the creation of the first stars.

The modern scientific perspective can remind us that our human history takes place in a context of forces vast beyond our everyday understanding. Events hidden deep in the past of the universe have contributed to the history of the land, to its very substance and nature. Our time and place are carriers of a past more profound than we imagine.

TECTONIC PLATES OF THE WORLD

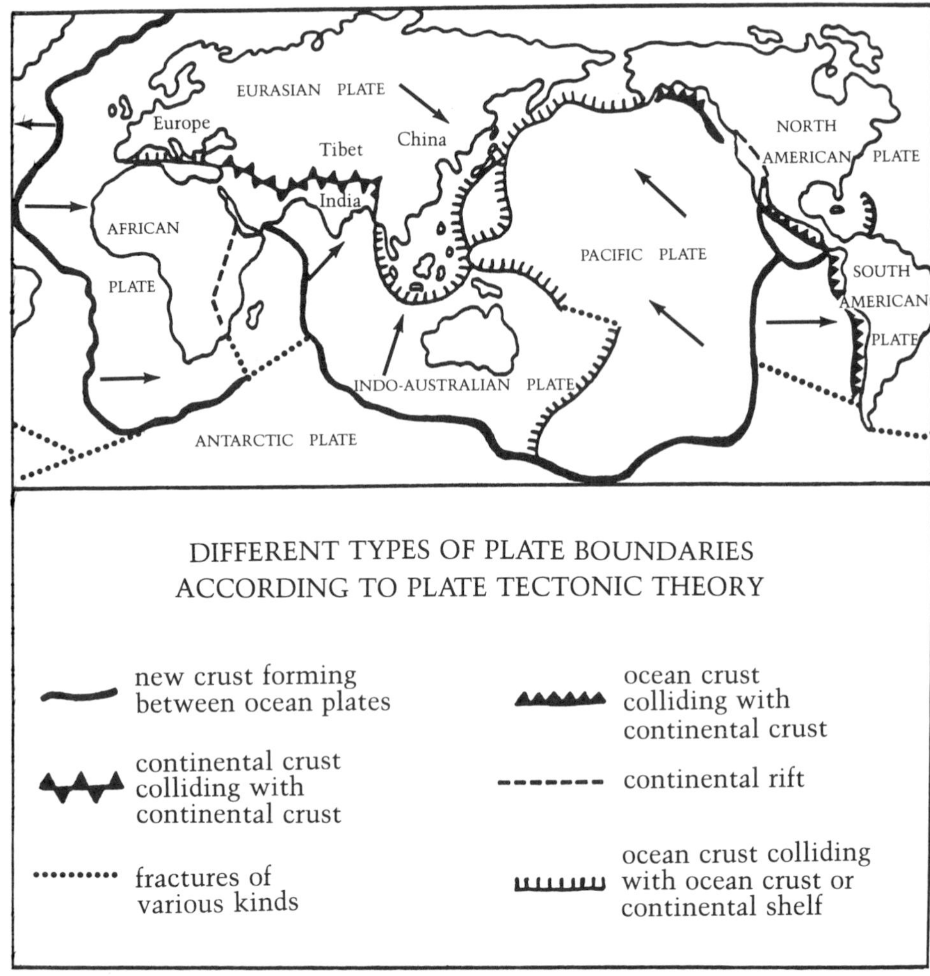

Recent geological discoveries indicate that the surface of the earth is divided into a number of separate plates that change positions, floating on top of the deeper deformable layers of the planet. The movements of these plates are thought to have created the mountains and oceans, while constantly shifting the position of the land. The modern arrangement of land is less than 100 million years old and is continuing to change as the plates move at the rate of two to eight centimeters per year (two to eight meters per century).

SEAFLOOR SPREADING

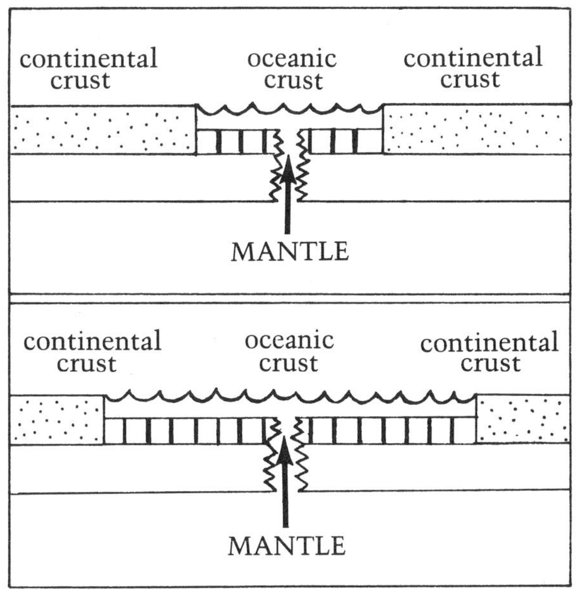

When a rift occurs in a plate carrying ocean floor, lava from the mantle rises up and hardens into new crust. Thus over time the seafloor spreads as new crust is created at the rift. The rock of this new crust takes on the characteristic magnetic polarity of the earth at the time of the rock's formation. Because this polarity shifts over time, bands of alternating polarity form on either side of the rift. This deep-sea record allows the dating of the ocean floor.

PLATE COLLISIONS

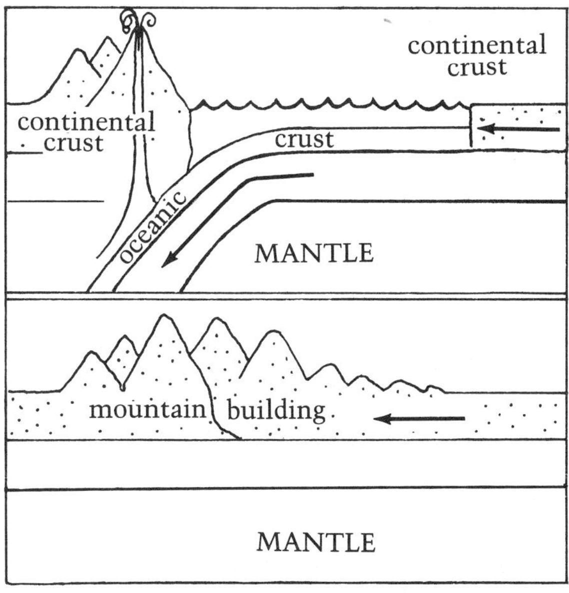

When a plate carrying oceanic crust collides with a plate carrying continental crust, the ocean floor bends beneath the continental plate, and is remelted in the deeper, hotter layers of the earth. This molten rock then rises, forming volcanoes along the edge of the continental plate. As the oceanic crust on a plate is consumed in this way, ocean between continents shrinks. This brings land masses into collision and builds mountain ranges as continental plates meet.

DEVELOPMENT OF SCIENTIFIC THEORIES

"Old Physics"	"New Physics"
Descartes' Physics Mechanical philosophy (1644)	Einstein's Special Theory of Relativity (1905)
Newton's Physics Law of gravity, mathematical models for physical processes (1687)	Atomic Theory: atom, nucleus, electrons (1913)
Laws of magnetic forces (1785)	Einstein's General Theory of Relativity (1915)
Electromagnetic Theory combines understanding of electricity and magnetism (1862)	Quantum Mechanics Theory (1924–26)
Electron Theory (1892)	Meson Theory (1932)
	Quarks postulated (1964)
	Charm Theory confirmed (1976)

SCIENTIFIC INVENTIONS AND DISCOVERIES

1609
Telescope

1654
Vacuum created

1706
Electrical conduction

1751
Lightning recognized as electricity

1800
Chemical battery

1821
Electric generator

1876
Telephone

1878
Electric lights

1896
Radioactivity discovered

1906
Vacuum tube

1922
Radio broadcasting

1932
Discovery of neutrons and antimatter

1942
Nuclear reactor

1943
Computers

1945
Atomic bomb

1952
High-energy accelerator: discovery of many subatomic particles follows

1953
Hydrogen bomb

1983
Discovery of W particle

GEOLOGIC TIMESCALE

PRECAMBRIAN	PALEOZOIC	MESOZOIC	CENOZOIC
14,000 mya Beginning of universe	575 mya Paleozoic Era opens	230 mya Mesozoic Era opens	65 mya Cenozoic Era opens
13,000 mya Galaxies form	550 mya Hard-shelled animals in seas	216 mya 1st primitive mammals	55 mya New mammals, primitive horses
11,000 mya Creation of many chemical elements	500 mya Earliest fossils from Tibet	200 mya Tibetan land rising out of sea	45 mya Himalayas forming
4550 mya Solar system forms	440 Ice Age	170 mya Petroleum forms in quantity	35 mya Early dogs and cats, New World monkeys
4450 mya Formation of earth's crust	425 mya 1st land plants	150 mya 1st birds	25 mya Climate cools, tropical vegetation declines, and grasslands spread
4000 mya 1st life on earth: primitive cells	380 mya Appalachian mountains forming	120 mya 1st modern flowering plants	19 mya Early antelopes
4000 mya Oldest continental rock	395 mya 1st insects	100 mya Gangs-ti-se, and gNyan-chen- thang-lha ranges built up	14 mya East Antarctic ice sheet grows
3500 mya Photosynthesis by primitive cells	370 mya 1st amphibians	Modern mammals developing	Volcanic activity worldwide
2800 mya Outer shell of earth hardens into moving plates	300 mya 1st reptiles	93 mya Sea level rises: widespread flooding of continents	7 mya Elephants evolve
2000 mya Atmosphere developing	Kun-luns, Urals, Pamirs, Altyn Tagh mountains forming	70 mya Earliest known primates	3.75 mya Ape-men develop
1500 mya Cells with organized nucleus	290 mya Ice Age	67 mya Catastrophic event? Worldwide layer of exotic iridium indicates cosmic meteorite impact	3 mya Modern horses develop
1000 mya Seaweed in oceans	260 mya Coal forms in quantity	Extinction of dinosaurs	Recent Ice Age begins
670 mya Jellyfish, worms in sea	235 mya 1st dinosaurs		2 mya Himalayas begin rapid rise
	1st flowering plants		

mya = millions of years ago

CHAPTER ONE

FORMATION OF THE TIBETAN PLATEAU

In the past thirty years researchers in geology, geography, meteorology, zoology, and botany have begun a thorough investigation of the Tibetan plateau. Seven large-scale multidisciplinary expeditions have taken place, and in 1980 two hundred scientists from seventeen different countries gathered in Beijing for a symposium on the plateau.

Recent investigations have inspired a number of theories about the formation of the Tibetan plateau. Though no single account is universally accepted, the general outlines of the plateau's development are beginning to emerge and can be related to known geologic events in the history of the planet.

Hundreds of millions of years ago, the Tibetan plateau was not in its present position and did not have its present shape. Most earth scientists now agree the plateau formed from several separate pieces of land that had very different histories. The land south of what is now the gTsang-po river appears to belong to the tectonic plate that carries India, while the northern regions of Tibet joined Asia independently. The following sections relate the ideas of earth scientists studying the ancient geography of the world, and summarize some of the reports presented at the 1980 symposium on the formation of the Tibetan plateau and the Himalayas.

ANCIENT CONTINENTS

The ancient stable rock beneath today's continents dates back between 2500 and 4000 million years ago. To these stable regions rock was added gradually through plate movements. The Tibetan plateau consists of at least thirteen layers, the oldest 660 million years old. By 550 million years ago, vast seas extended across the north and south poles, and the land masses that would later make up Asia were scattered in separate pieces near the equator. Between these ancient continents lay shallow seas, which often invaded the land. North America, part of Europe, Siberia, China, and Kazakh (western Central Asia) were widely separated by water. But other land masses were grouped more closely together: Australia, Antarctica, Africa, South America, Arabia, southern Europe, India, and the southern part of Tibet. This cluster has been named Gondwanaland by earth scientists.

About 450–430 million years ago, India and the southern portions of Tibet, together with the rest of Gondwanaland, drifted far to the south. Parts of Africa and Australia lay over the south pole, and the Sahara was buried under a blanket of ice and glaciers. Southern Tibet seems to have been well south of the equator, for fossils in ancient rock layers indicate the land at that time was covered with cold water. Reconstructions of locations of ancient land masses published in 1979 suggest that southern Tibet was located about the latitude of modern Australia.

A SUPERCONTINENT FORMS

Between 400–300 million years ago, the primitive continents began to join together, forming a supercontinent that has been named Pangaea, "All-Earth." It is believed that several earlier supercontinents had existed in the very ancient past: 2800 million years ago, 1500 million years ago, and 800 million years ago. It was during the formation and consequent breakup of the fourth supercontinent of Pangaea that the Tibetan plateau took shape. To envision the creation of the plateau in its global setting, it may be helpful to consider the major events in the history of Pangaea.

A reconstruction of Pangaea was worked out in 1970, and discussions and diagrams of the formation and breakup of Pangaea can now be found in introductory geology textbooks and general encyclopedias. The first plate collision that began to build Pangaea was between northern Europe and North America, raising the Appalachian mountains and the ranges in northwestern Europe. Then massive Gondwanaland and new "EurAmerica" joined together so that South America and Africa lay close to North America and western Europe. India, Australia, and Antarctica were still neighbors on the other side of Africa. Meanwhile Asia was coming together, as Siberia and Kazakh joined with north China. The newly formed Asia then collided with "EurAmerica" about 300–250 million years ago, raising the Ural mountains in between Siberia and Europe. This combined mass of Europe, America, and Asia has been named Laurasia.

Most of the land masses on the earth were now connected to each other, but Pangaea's northern portion, Laurasia, and its southern portion, Gondwanaland, were partly separated by a body of water known as the Tethys Sea. As continents joined together, mountain-building took place all across the world. About the same time, a great ice age occurred in the southern hemisphere, the sea level dropped, and a magnetic reversal of the poles took place. The possible interconnections among all these events are being investigated by earth scientists.

Almost as soon as the supercontinent had formed, it began to split up. About 200 million years ago, the continents started to drift toward their modern positions. North America was first to split away, heading west from Africa and Europe. This movement began to open the Atlantic Ocean and eventually would build the Rockies in North America. Australia and Antarctica headed farther south, while India first drifted east away from the coast of Africa and then about 100 million years ago started north. The Indian plate, bearing southern Tibet, now headed toward Asia.

As India approached Asia and Africa rotated north toward Europe, the Tethys Sea gradually disappeared, leaving only the Black Sea and the Caspian Sea as surviving remnants. Africa's northward movement also began to thrust up the Alps in Europe about 100 million years ago, while India's northward drift and subsequent collision with Asia raised the chain of mountains called the Himalayas.

CREATION OF THE TIBETAN PLATEAU

The northern parts of Tibet appear to have joined Asia earlier than southern Tibet, which belonged to the Indian plate that collided with Asia some 45 million years ago. Recent evidence indicates that Tibet's most northerly ranges are the oldest, those across the northern plain of Byang-thang are younger, those in central Tibet are younger still, and the Himalayas are the youngest of all. Such a chronological order suggests to some researchers that land was gradually added to Asia to form the Tibetan plateau, the last piece arriving with India.

These northern additions may have been blocks that broke away from Gondwanaland, islands in the Tethys Sea, or small land masses somewhere between Asia and Gondwanaland. The history of these northerly regions is still uncertain, but reports from the 1980 symposium and papers published in recent Western journals offer this interesting scenario:

About 300 million years ago, India still lay in the southern hemisphere, along the east coast of Africa. It was separated by the Tethys Sea from north China and Tarim, the block of land later to become the site of Khotan, north of Tibet. The Kun-lun range that today runs between the Tarim basin and the northern edge of the Tibetan plateau is thought to have formed by this time.

Byang-thang seems to have joined to Asia by 200 million years ago when Pangaea began to break apart. Across the northern edge of the plateau today extends a region that appears to be a connection zone along Litian and Margai Caka lakes, south of the Kokoshili mountains, and along the 'Bri-chu. About this same time, south China and Indochina were joining Asia. Byang-thang is thought to have been connected to the land southwest of the Red river region in Indochina (northern Vietnam, Cambodia, and Laos), and these regions may have joined Asia together.

Another connection zone appears to run between Byang-thang and central Tibet, from Pang-gong lake in the far west, across Byang-thang, and along the Nag-chu river (Nu Jiang) in the east. Thus, the region of central Tibet south of the Nag-chu river and north of the gTsang-po seems to have joined Asia next. This land may also have been connected to Thailand and the Malay peninsula, and these regions may have moved together. The most recent surveys indicate that fossils south of the Nu Jiang connection zone are typical of the Indian subcontinent, suggesting the region was associated with the Indian plate.

As south China, Indochina, Byang-thang, and central Tibet joined Asia, the series of collisions and additions began to raise northern Tibet out of the water. The Kun-luns and the region east of the 'Bri-chu rose above sea level, followed by Byang-thang north of Gangs-ti-se, as well as the region north and east of the Nag-chu river: the modern area of Chab-mdo, Ri-bo-che, Tsha-ba-rong, and dMar-khams. The plate motions are also thought to have produced great folding in older mountain ranges farther north, such as the Kun-luns and the Bayankara mountains.

THE BIRTH OF THE HIMALAYAS

The joining of the far southern part of Tibet to Asia is better understood, and discussions of India's northward drift can be found in encyclopedias as well as in geology texts. The 1980 symposium presented new details on this collision, giving a variety of hypotheses to explain the uplifting of the plateau.

By 100 million years ago, the plate carrying India had begun to drift north from its location alongside Africa as rifts formed in Gondwanaland. This opening between India and Africa eventually became the Indian Ocean. As India moved north, ocean crust beneath the Tethys Sea was first to press against Asia. Sliding down beneath the continental land along the line now formed by the gTsang-po river, the ocean crust was remelted, and the sea began to shrink. The northward pressure and the remelting of crust created intense agitation. New volcanoes erupted in Gangs-ti-se and gNyan-chen-thang-lha, building these ranges higher. By 50 million years ago, the bulk of Mount Ti-se was built up. About this time the Rockies were forming in America and the Caucasus formed in Russia. North America was just now splitting away from Europe as the Atlantic Ocean continued to widen.

By 45–40 million years ago, the entire crust of the Tethys Sea had descended beneath the Asian continent, closing the sea completely. The Indian plate carrying southern Tibet now collided with central Tibet along the zone of today's Indus and gTsang-po rivers. As land masses met head-on, the Himalayas began rising out of the sea. The last oceanic crust and sedimentary rock overlying ocean floor, as well as volcanic islands that had formed between the plates, were compressed and squeezed into huge folds that slid south over the Indian plate. The Himalayas continued to grow by a southward extension that occurred in several stages.

As India continued to press north, the crust fractured, and about 20 million years ago, large faults developed in the new Himalayas. Along a fault 2000 kilometers long, known as the Main Central Fault, thrusting movements pushed older rock layers up over younger layers. This great northward pressure contracted the plateau, thickening its crust, and exerted force on all the older mountains, creating more folding and thrusting these ranges up even higher. Gangs-ti-se greatly increased at this time. Volcanoes began erupting north of the collision zone, gradually forming farther and farther north, until they arose along the Kun-lun mountains. This series of volcanoes, which are now extinct, can be seen south of Muztagh Ata peak in the Kun-luns and south of Margai Caka salt lake in Byang-thang.

THE UPLIFT OF THE HIMALAYAS

Though the Himalayas and the plateau had formed by 20 million years ago, it took many millions of years more before the mountains and plateau were uplifted to their present elevation.

Before this uplift began, the Himalayas were perhaps no higher than the Alps or Rockies are today, while the plateau was a low-lying plain, perhaps no higher than the Indian subcontinent is today.

The causes of the uplifting of the Himalayas and the Tibetan plateau to create a three-mile-high platform are not yet completely understood, and there are different opinions about their elevation at various times. According to some earth scientists, between 10 and 5 million years ago, the plateau had reached only 3000 feet (1000 meters) above sea level, while the Himalayas were about 9000 feet (3000 meters) above sea level. Other researchers point out that some rivers on the plateau cross the mountain ranges, and suggest that until 2 million years ago, the Tibetan plateau might actually have been higher than the Himalayas. Major rivers such as the Indus and gTsang-po managed to cut their way through the Himalayas, incising deep gorges as fast as the new mountains were rising.

Rapid uplift of the plateau seems to have begun about 2 million years ago, as the Indian plate thrust down at the southern foot of the Himalayas along a fault called the Main Boundary Fault. Compressed between the Tarim mountain system in the north and the Indian subcontinent in the south, both the plateau and its mountain ranges increased in elevation. Tibet was probably rocked by frequent earthquakes and tremors during this period, which coincided with the most recent ice age. A million years ago the plateau may have been 9000 feet high (3000 meters), and the Himalayas may have reached 15,000 feet (4500 meters). By 10,000 years ago, the plateau may have reached over 15,000 feet in elevation (4700 meters), and the Himalayas may have been over 19,000 feet (6000 meters), about the height of the Kokoshili mountains across northern Byang-thang today.

The great height of the plateau seems related to the intense pressure exerted by India, which appears to have forced the separate pieces of the plateau to overlap one another from north to south. The extreme uplifting may also be related to the sliding of the Indian plate under the plateau, though how far the plate might extend is debated. The crust beneath the plateau has been measured in some places at 65–70 kilometers, which is about twice as thick as the average continental crust, suggesting that the Indian plate may extend some distance beneath the plateau.

Another model for the tremendous uplift of the Himalayas proposes that the Indian plate fractured and that "splinters" of plate wedged successively under each other. The splinter model may also explain the tilting of mountain ranges such as the Himalayas, so that the south slopes are steep, but the north slopes are gentle. Other ranges that exhibit tilting, such as Gangs-ti-se and Kokoshili, may have formed in similar ways as pieces of land wedged beneath each other.

DEVELOPMENT OF RIVERS AND LAKES

The collision between India and Tibet and the continued northeastward pressure on the plateau may also be responsible for the unusual arrangement of rivers in eastern Tibet. Through the deep gorges of Khams flow three rivers in parallel courses that in some places are only 70 kilometers apart: the Nag-chu–dNgul-chu (Salween), the rDza-chu (Mekong), and the 'Bri-chu (Yangtze). Northeastward movement of the plateau may have squeezed the rivers together, moving them as much as 500 kilometers from an older, more westward position.

This same movement may explain the peculiar tributary system along the gTsang-po. The sKyid-chu and other small rivers that enter the gTsang-po approach it at an upstream angle, flowing to the west, while the gTsang-po itself flows to the east. This is rather uncommon—tributaries usually flow the same direction as the main stream. Perhaps the gTsang-po formerly flowed west instead of east. But it is possible that these tributaries used to be much farther west, flowing east to empty into the gTsang-po. The northeastward squeezing of the plateau may have repositioned these streams so far to the east that to enter the gTsang-po from their new positions, they must flow west.

As the plateau began to increase in elevation, water filled the basins that had formed from faults and rifts in the crust and from hollows that had been left between rising mountains. Many inland lakes arose this way. Other lakes were created as a result of blockage of rivers by volcanic debris or by silting and rock slides. Still other lakes formed as glaciers retreated, leaving moraines that blocked glacial valleys and held meltwater.

Lakes began drying out, however, as the plateau rose into drier air and the Himalayas became high enough to block moist winds from the south. Rings of strand lines can be seen around most lakes today, marking older, more extensive shorelines. Many lakes had been connected to rivers, but as the water level fell, strings of drying lakes were left behind. The lakes south of the Kun-luns may have belonged to the 'Bri-chu, and lakes between Pang-gong lake and the source of the Nag-chu may have been part of one river system. In the past Zigetang lake opened into Zilling lake, little lakes around Pang-gong lake were part of Pang-gong, and Ma-pham and Lag-ngar formed one larger lake.

The rising of the plateau not only dried out inland bodies of water, but converted fresh water lakes to salt water ones. Minerals and salts from natural erosion of rock are always present in lakes, so as lake water evaporated, the water that remained grew increasingly saline. The northern part of the plateau seems to have suffered most from this process, and great salt deposits are found across northern Byang-thang around drying lakes. The lakes in southern Byang-thang are large, and some are freshwater, while the southeast has numerous fresh-water lakes.

The formation and evolution of the mountains, rivers, and lakes on the Tibetan plateau, as well as the development of the plateau itself, are of great interest to scientists from all over the world. Research into the history of the youngest mountains and the highest plateau on the planet is expected to yield valuable knowledge about the tectonic processes that continue to shape the face of the earth.

FORMATION OF THE PLATEAU

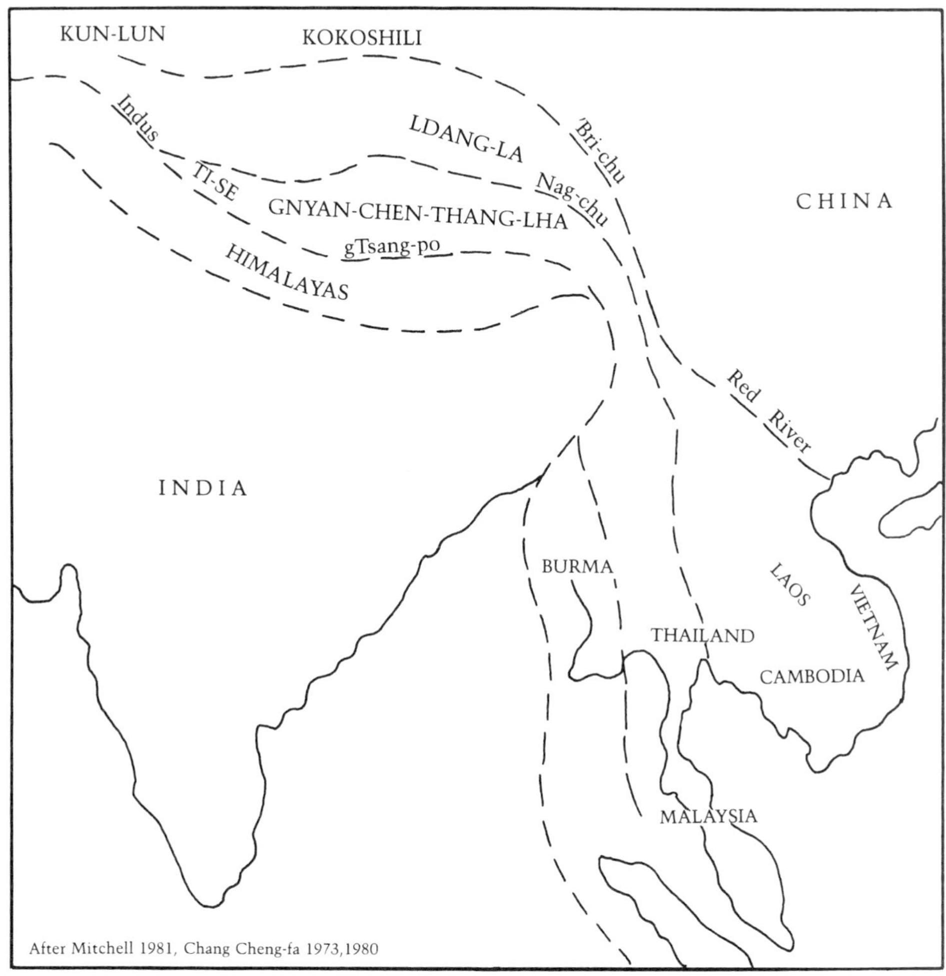

According to one current theory, the Tibetan plateau was assembled from several pieces of land that were added to Asia at different times. This theory is supported by the decreasing age of mountain ranges from north to south, and by the presence of what appear to be suture zones running across the plateau. About 200 million years ago, two distinct land masses joined Asia, one after the other. Both moved together with regions of Southeast Asia. A third and final piece arrived 45 million years ago with the Indian plate.

SHIFTING CONTINENTS

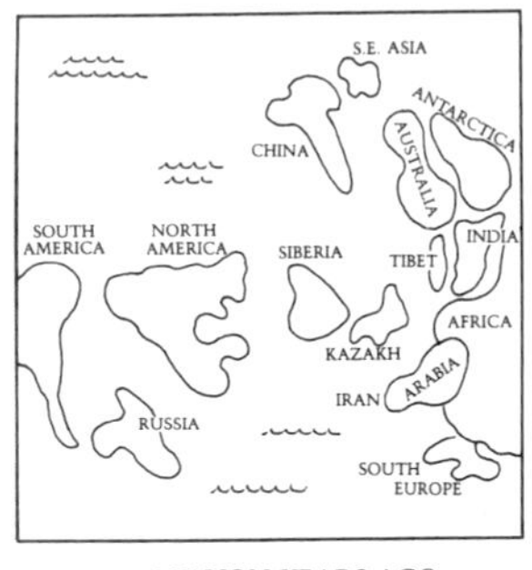

550 MILLION YEARS AGO

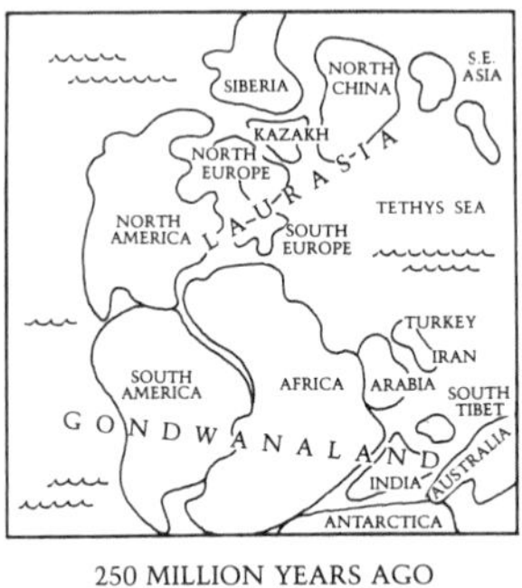

250 MILLION YEARS AGO

135 MILLION YEARS AGO

During the long ages of geologic history, the continents have shifted positions many times. Over hundreds of millions of years, the ancient land masses drifted closer together and eventually formed a supercontinent known as Pangaea. The southern regions have been named Gondwanaland, while the more northerly regions are called Laurasia. Soon after Pangaea formed, it began to split apart, as the continents drifted toward their modern locations.

INDIA COLLIDES WITH ASIA

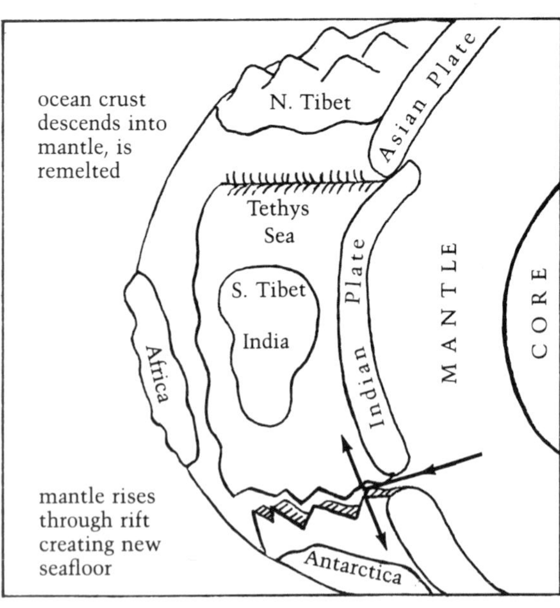

When a rift formed between the Indian plate and Africa and Australia, molten rock erupted through the rift. This created new seafloor, which spread as the Indian plate moved north; the old seafloor descended into deeper layers of the earth along a trench at the northern edge of the Indian plate, closing the Tethys sea between Asia and India. As the plate progressed north, ocean floor was completely remelted, bringing the land masses on the Indian and Asian plates into collision.

EVENTS OF THE ANCIENT PAST

MOVEMENTS OF CONTINENTS	WORLDWIDE SHIFTS	MOVEMENTS OF CONTINENTS	WORLDWIDE SHIFTS
550 mya North America, North Europe, Siberia, China, Kazakh scattered across seas near equator	670 mya Severe Ice Age destroys marine plant life	210 mya North America leaves Africa	210 mya Canadian crater indicates cosmic impact
Gondwanaland: Australia, Antarctica, Africa, Arabia, India, South Tibet grouped together	440 mya Ice Age: mass extinctions of trilobites and early fish	200 mya Southeast Asia and most of Tibet join Asia	183 mya Russian crater indicates cosmic impact
450–400 mya Europe and North America collide	430 mya Magnetic reversal of poles	140 mya India separates from Africa	182 mya Halt in increasing diversity of marine animals
450–430 mya Gondwanaland moving south, Australia and Africa over south pole	370 mya Swedish and Canadian craters indicate cosmic impacts	Eurasia and Africa move apart	85 mya Era of frequent magnetic reversals: 177 shifts to present with highest frequencies at mya: 72, 54, 42, 24, 15
350–280 mya EurAmerica collides with Gondwanaland	370 mya Widespread destruction of coral, trilobites, and brachiopods	140 mya Korea joins Asia	67 mya American and Russian craters; worldwide layer of exotic iridium indicates cosmic impact
350–280 mya Kazakh, Tarim, N. China, and Siberia unite	300 mya Magnetic reversal of poles	130 mya South America separates from Africa	67 mya Sudden extinction of dinosaurs, all land animals over 25 kilograms; many marine forms disappear
300–260 mya Asia collides with EurAmerica forming Laurasia	290 mya Ice Age	60 mya North America separates from Europe	38 mya Canadian and Russian craters; iridium deposits indicate cosmic impact
260–220 mya Pangaea complete	245 mya 96% of marine animal species disappear; collision with comet conjectured	50 mya Australia leaves Antarctica	15 mya German crater indicates cosmic impact
	216 mya Mass extinction of marine reptiles, certain sponges, and mammal-like land reptiles	45 mya India collides with Asia	
		30 mya Japan moves away from Asia	
		20 mya Arabia joins Eurasia	
		3 mya North and South America join	

mya = millions of years ago

CHAPTER TWO

MODERN LANDFORMS

The formation of the Himalayas and the elevation of the Tibetan plateau are the most recent large-scale mountain building events in the history of the planet, and continue even today. As India moves north, pressing against Tibet, the plateau and Himalayas are pushed up. Published calculations state that the rate of increase varies from one to five inches per year.

Other evidence of continuing crustal movements in Tibet includes deep fault zones, severe earthquakes, and unusual geothermal events, all topics discussed at the 1980 symposium. The continuing pressure between India and Tibet is thought to be related to earthquakes and plate movements throughout Asia, and studies of the structure of the plateau may shed light on plate tectonics in this half of the world. The following sections report some of the findings presented at the symposium.

FAULT ZONES AND EARTHQUAKES

As a result of the collision and the continuing pressure from India, the Tibetan plateau may be undergoing an extension or squeezing in an east-west direction, which is stretching the surface crust. This stress may be related to the north-south faults that run across parts of Tibet, especially south of the zone between Pang-gong lake and the Nag-chu river.

Maps based on satellite photos of the plateau reveal large faults in the Himalayas as well as deep fault zones across northern Tibet, which turn south and east to stretch down through Khams. Deep faults are often found along what appear to be old plate boundaries. Where plates are sliding past each other, faults are called "strike-slip" or "transform" faults. These are thought to exist in the Kun-luns; in the Karakorums; in the Altyn Tagh range; along the She-chu and Nyag-chu, extending into eastern Yunnan (Xianshui-Anning river-east Yunnan fault zone); along the 'Bri-chu and the Hong He, extending into northern Vietnam (the Jinsha river-Red river fault zone); and in the Longmen fault zone south of Lanzhou (Lan-chou). "Thrust" faults, where plates are pressing against each other, are thought to exist in many places in Tibet and in the Himalayas where the Indian plate presses against Asia.

Earthquakes most often occur along fault zones where plates move against each other. Earthquakes on the plateau are often shallow, while quakes in the mountain systems surrounding the plateau usually originate deeper in the crust. Since the turn of the century, six great quakes have taken place in and around Tibet: to the far west beyond the Indus in the Kashmir region in 1905; to the south in the Himalayas in Nepal in 1934; to the southeast, beyond the bend of the gTsang-po in 1950; to the northeast near Lanzhou (Lan-chou) and Tsaidam in 1920 and 1927; and near lHa-sa in central Tibet in 1951. Five of the six were in the mountains surrounding the plateau, and only one was in the interior of the plateau. The most active zones seem to be in the Himalayas and on the east and west borders of the plateau.

GEOTHERMAL ACTIVITY

Continuing activity of the crust is also indicated by the presence of a geothermal belt, a zone where the earth's internal heat is released at the surface. Over 600 geothermal areas such as hot springs, geysers, and sites of hydrothermal explosions have been recorded. Most are south of Gangs-ti-se and gNyan-chen-thang-lha and north of the Himalayas. Hot springs are especially prevalent along either side of the gTsang-po river. These springs are not associated with volcanic activity, as is usually the case in other parts of the world. No active volcanoes have been found in Tibet, although along the Kun-lun mountains and scattered across northern Byang-thang are volcanoes that were active as recently as 2 million years ago.

The presence of hot springs and geysers points to ongoing plate movements and increased temperatures beneath Tibet. Magnetic studies and seismic data also suggest that molten materials lie below the surface. This geothermal region, corresponding to the collision zone between India and Tibet, is part of a belt that extends through the plateau, west into Iran, Turkey, Greece, and Italy, and east into Yunnan and Burma. The increased temperatures beneath the surface across this zone are apparently related to the plate movements that built the Himalayas, closed the Tethys Sea, and moved India and Africa north.

Efforts to collect scientific data continue. Geologists hope that study of the unusual structure of the crust of the plateau and the deeper layers beneath it will reveal more about mountain-building events. In addition to theoretical and historical concerns, scientists are interested in developing geothermal sources for electric power, and a number of power plants are now under construction at Yang-ba-can (Yangbacain) and other sites located in the geothermal belt.

MOUNTAIN RANGES

While investigation of the structure of the deep layers of the plateau is very recent, the surface features of the Tibetan highland have been studied and mapped over the past century or so, beginning with early explorers and continuing with recent scientific expeditions and satellite photography. The following information has been collected from maps based on satellite photos and from materials published by the U.S.S.R. Academy of Science. The plateau is criss-crossed by twelve sets of ranges, each with its own distinctive character.

KUN-LUNS

The Kun-luns are a group of S-shaped ranges forming the northern border of the Tibetan plateau, separating Byang-thang from the Tarim basin and running along the southern edge of the Tsaidam basin. The longest mountain system in Asia, the Kun-luns stretch nearly 1500 miles from the western end of the plateau, where they are separated from the Karakorum range by the Yarkand river, to the Koko Nor region in the east. Northern slopes are more rugged than southern slopes that meet the high plateau. In the west the Kun-luns consists of parallel chains of mountains bending around the Tarim basin, with many peaks rising 20,000–22,000 feet (6000–7000 meters). Central and eastern ranges in the Kun-luns are less elevated, generally between 17,000 and 19,000 feet (5000–6000 meters). The highest peak is Ulug Muztagh, rising 25,476 feet (7723 meters).

Beyond the northern flanks of the Kun-luns, the altitude drops steeply at the rim of the Tarim basin. Oases such as Hotan (Khotan), Yutian (Keriya), and Qarqan (Charchan) are located only 3000–5000 feet (900–1500 meters) above sea level. The rivers of the western Kun-luns drain north into this basin and are fed mostly by glaciers in the western peaks. Untouched by the southwestern monsoon from the Indian Ocean, the landscape in these mountains is deserty and arid. The snowline lies at 17,000–19,000 feet (4500–5100 meters), much higher than the snowline in the Himalayas far to the south.

ALTYN TAGH

Branching off from the Kun-luns is the Altyn Tagh range, which forms the northwestern border of Tsaidam. In the west, this range rises above 19,000 feet (5700 meters); the slopes are rocky and rugged, and a few peaks have glaciers. As Altyn Tagh stretches across the northern border of Tsaidam, its elevation decreases to about 15,000 feet (4500 meters). Extending over 500 miles, this range is continued in the east by the Nan Shan mountains.

NAN SHAN

The Nan Shan range forms the northeastern border of Tsaidam, separating it from the Gobi desert. These mountains run parallel to the eastern Kun-luns at about 17,000 feet (5100 meters). While the Altyn Tagh is extremely dry, the valleys between the narrow Nan Shan ranges carry rivers from glaciated peaks. Thus grazing land can be found up to 14,000 feet. From the northern slopes the Ruo Shui (Kan-chou river) flows into Mongolia as the Etzingol; other streams drain into Central Asia and into rivers running into the Pacific. The ranges south of Koko Nor lake are often considered part of the Nan Shan ranges. The Koko Nor depression, the lower flatland at 10,000 feet where Koko Nor lake lies, can be viewed as a depression within this range. The Nan Shan ranges contain many deposits of iron, chromium, copper, lead, zinc, gold, and coal.

KOKOSHILI

The Kokoshili range is often considered part of the Kun-lun ranges although its history and development appear to have been distinct. These mountains cross the northern central part of Byang-thang, parallel to the eastern Kun-luns. Many peaks rise over 20,000 feet (6000 meters), yet due to intense weathering, the peaks are often dome shaped and flat rather than rugged. In the west, snow-covered peaks rise 21,000 feet (6300 meters). Fairly large lakes abound in the Kokoshili range, filling depressions between the mountains.

BAYANKARA

The Bayankara range is also often considered part of the Kun-luns. It runs in a northwest-southeast direction, south of the lakes sKya-rengs-mtsho and sNgo-rengs-mtsho (Oring and Tsaring), and parallel to the A-mnyes-rma-chen range. In between rMa-chen and the Bayankaras stretches a foothill region at an elevation of about 14,000 feet (4200 meters). The Bayankaras rise about 16,000–17,000 feet (4800–5100 meters). The northwestern Bayankaras are the source of the rMa-chu river (Huang Ho). Eastern extensions of this range include the magnificent snow-clad gNyan-po-g·yu-rtse, a sacred peak rising 16,490 feet (4947 meters) and flanked by five little lakes.

A-MNYES-RMA-CHEN

The A-mnyes-rma-chen range also runs in a northwest-southeast direction, extending into the knee-shaped bend of the rMa-chu river known as the khug-pa. The highest elevations, over 17,000 feet (5100 meters), are in the northwest, while in the east elevations drop to 14,000 feet (4200 meters). Three peaks rise together to form the pinnacle of the range in the northwest. The northernmost peak, rising 20,731 feet (6282 meters) is known as dGra-'dul-rlung-shog. In the center stands the peak known particularly as A-mnyes-rma-chen, while the southern peak is called sPyan-ras-gzigs. A-mnyes-rma-chen is famous throughout Tibet as the home of the god rMa-chen-spom-ra, and pilgrims circumambulate the western end of the range along ancient trails.

LDANG-LA

This range also runs northwest-southeast, separating the northeastern part of the plateau from the more central regions of dBus-gTsang. Old caravan trails ran through the passes in this range on their way to and from central Tibet. Many peaks in the lDang-la range rise over 19,000 feet (5700 meters), some reaching over 21,000 feet (6300 meters). Here is the source of the 'Bri-chu (Yangtze). Deposits of iron ore, hard coal, graphite, asbestos, and soapstone are found throughout the lDang-la mountains.

KHAMS-SGANG-DRUG

Six mountain ranges lie close together in Khams, making nearly the whole of this region mountainous terrain. Southeastern Tibet is often called the River Gorge Country because of the deep river valleys, steep slopes, and sharp, narrow ridges. This region is often referred to as the Land of Six Ranges and Four Rivers.

Zab-mo-sgang: This range, also known as Zal-mo-sgang, lies between the Ngom-chu and the upper Nyag-chu at elevations of 16,000–18,000 feet (4800–5400 meters).

Tsha-ba-sgang: Tsha-ba-sgang extends between the dNgul-chu (Salween) and the rDza-chu (Mekong) at elevations of 17,000–19,000 feet (5100–5700 meters). Just to the southeast is the Kha-ba-dkar range whose highest peak, Kha-ba-dkar-po, reaches a height of 22,340 feet (6702 meters).

sPo-bo-sgang: sPo-bo-sgang rises north of the bend of the gTsang-po river as it turns south to flow through the Himalayas into India. Many of its peaks rise 19,000–21,000 feet (5700–6300 meters). In the bend of the gTsang-po lies gNam-lcags-bar-ba, a magnificent peak rising 25,595 feet (7756 meters).

sMar-khams-sgang: sMar-khams-sgang, also known as Ningjing Shan, lies between the rDza-chu (Mekong) and the 'Bri-chu (Yangtze) in southern Khams. Average elevations are 17,000–19,000 feet (5100–5700 meters).

Mi-nyag-sgang: Mi-nyag-sgang is located between the lower rGyal-mo-dngul-chu (Dadu He) and the She-chu and lower Nyag-chu (Yalong) at elevations of 16,000–18,000 feet (4800–5400 meters). Just thirty miles south of Dar-rtse-mdo is Mi-nyag-gangs-dkar (Minya Konka), a famous snow-capped mountain rising 25,047 feet (7756 meters).

gYar-mo-sgang: gYar-mo-sgang extends between the 'Bri-chu (Yangtze) and the lower Nyag-chu (Yalong) at elevations of 17,000–19,000 feet (5100–5700 meters).

TRANSHIMALAYA

gNyan-chen-thang-lha, Gangs-ti-se, and mNga'-ris-gangs-ri are often grouped together as the Transhimalayan ranges. These mountains form the watershed between the Indian Ocean and landlocked northern Tibet, and establish a natural boundary between Byang-thang and southern Tibetan environments.

GNYAN-CHEN-THANG-LHA

gNyan-chen-thang-lha lies north of the gTsang-po river and south of the large lake gNam-mtsho. It extends northeast and then curves south under the Nag-chu (Nu Jiang). North of this range, the arid Byang-thang begins, while to the south is a more moist region of valleys where tributaries of the gTsang-po flow. Elevation is 19,000–20,000 feet (5700–6000 meters). The highest peak is the holy mountain of gNyan-chen-thang-lha, rising 23,390 feet (7080 meters). Like the Himalayas, this range exhibits steep and rugged south slopes with more gentle northern slopes.

GANGS-TI-SE

Also known as the Kailash range and Kangtissushan, the Gangs-ti-se mountains run along the north side of the gTsang-po and stretch into mNga'-ris, the western region of Tibet. The south slopes are sharp and rugged while the north slopes are more rounded and sloping. Elevation reaches 20,000 feet (6000 meters), with Ti-se (Mount Kailash, Kailāśa) rising 22,156 feet (6714 meters). This famous holy mountain is revered in several religious traditions and has been circumambulated by pilgrims for many centuries. Just southwest of Ti-se are two equally famous lakes, Rakas Tal (Lag-ngar-mtsho) and Manasarowar (Ma-pham-mtsho), which are separated by an isthmus. Within forty miles of these lakes and the holy peak are the sources of some of the greatest rivers of Asia: the gTsang-po, the Sutlej, the Ganges, and the Indus. In Eastern traditions this region is sometimes called the "Center of the World."

MNGA'-RIS-GANGS-RI

This range lies west of the Gangs-ti-se range in northern mNga'-ris. Peaks often reach 19,000–21,000 feet (5700–6300 meters), with the snow-capped mountain, mNga'-ris-gangs-ri, rising 21,160 feet (6348 meters) at the northwest end of the range. Towards the southeast near Ngang-la-ring lake another snow peak rises 21,430 feet (6429 meters). This range differs from the other two making up the Transhimalayas in being much less rugged and more similar in relief to Byang-thang ranges such as Kokoshili.

THE HIMALAYAS: ABODE OF SNOW

This great range circles the southern edge of the Tibetan plateau and turns north, running along the Tibet-India border up to Kashmir where the Indus river separates it from the Karakorum range. The highest mountain range in the world, the Himalayas, meaning Abode of Snow in Sanskrit, have more than thirty peaks over 25,000 feet (8000 meters). These ranges contain valuable minerals such as antimony, arsenic, molybdenum, copper, zinc, and lead. The south side of the Himalayas is very rugged while the north side that meets the high plateau is more gentle. On the south side, the snowline lies at 15,500 feet (4600 meters), but rises higher and higher progressing towards the northern slopes. The north side is drained by tributaries of the gTsang-po river, which begin in Himalayan glaciers.

Three longitudinal zones can be distinguished in the Himalayas: the Great Himalayas, snow-covered peaks above 20,000 feet; the Lesser Himalayas, which include the mountains of Kashmir and Nepal; and the Outer Himalayas, made up of the Siwalik range in India.

The following peaks stand out from west to east: Gangotri glacier at 23,421 feet (6726 meters); Kamet at 25,595 feet (7756 meters); Nanda Devi at 25,796 feet (7817 meters); Gang-lung at 21,080 feet (6324 meters); Kanjiroba at 21,640 feet (6492 meters); Dhaulagiri at 25,967 feet (8172 meters); Anna Purna at 26,657 feet (8078 meters); Gosainthan (Xixibangma) at 26,440 feet (8012 meters); Gung-thang-la at 23,990 feet (7197 meters); Jo-mo-gangs-dkar (Mt. Everest, Qomolangma) at 29,198 feet (8848 meters);

Gangs-chen-mdzod-lnga (Kanchenjunga) at 28,373 feet (8598 meters); Jomo-lha-ri at 24,133 feet (7313 meters); Phu-la-ha-ri at 24,700 feet (7410 meters); and gNam-lcags-bar-ba at 26,595 feet (7756 meters).

Other well-known mountain peaks in and around Tibet include rTsa-ri, a holy mountain south of the gTsang-po river rising 18,813 feet (5644 meters); Bya-rkang southeast of Tibet near the Dali lake region; Yar-lha-sham-po at the south end of the Yar-lung river, a mountain sacred to the early kings of Tibet; and Ha-se, also known as gNod-sbyin-gangs-bzang, between rGyal-rtse and Yar-'brog in central Tibet, rising 22,300 feet (6690 meters).

LAKES AND RIVERS

The Tibetan plateau is dotted with many spectacular lakes, some saline and some fresh. Byang-thang has numerous saline lakes of all sizes and shapes. Among the largest lakes in Tibet is Koko Nor in A-mdo, over 60 miles across and larger than Lake Michigan in North America. The lakes Oring and Tsaring are each about 25 miles wide. In central Tibet gNam-tsho measures about 50 miles across as does Zilling lake, while Dang-ra is about the same distance long, though not so wide. Yar-'brog's several petal-shaped extensions are 40 miles across. In the far west Lake Ma-pham is about 15 miles across and over 240 feet deep, while its companion Lag-ngar is about 15 miles long. Lake Pang-gong in the northwest, a long L-shaped lake, has one arm running about 30 miles from north to south, and the other extending about 70 miles east to west.

The lakes on the plateau are an important source of minerals such as rock salt, mirabilite, gypsum, borax, magnesium, potassium, lithium, rubidium, cesium, strontium, uranium, and thorium. Though rich in minerals, the lakes are filled with extremely clear water, for very little soil muddies them. The thin air at such high elevations allows strong sunlight to penetrate deeply into the clear water, creating the bluest lakes in the world.

The Tibetan highlands are a major source of water for all of Asia, and many of the world's longest rivers have their origin in Tibet. The famous rivers of eastern Tibet originate in the interior of the plateau while the four great rivers near Kailāśa in the west arise in the mountains bordering the plateau.

RMA-CHU

The rMa-chu (Ma Qu) arises beyond sKya-rengs and sNgo-rengs (Oring and Tsaring lakes) in the Bayankaras, flows east through 'Gu-log, and makes a huge "hairpin turn" called the "knee" (khug-pa) to flow back toward the west. Another bend is made at a right angle as the rMa-chu turns to flow north into A-mdo. Again it turns east toward China where it is called the Huang Ho or Yellow river and flows to the Pacific. Its 2903 mile course makes it the world's sixth longest river.

NYAG-CHU

The Nyag-chu (Za Qu) originates northeast of Khyer-dgun-mdo in the Bayankara mountains. Along its upper course it is known as the rDza-chu, not to be confused with the rDza-chu farther southwest. It gives this name to the surrounding region south of 'Gu-log, which is known as rDza-chu-kha. In its middle course, it is known as the Nyag-chu and lends its name to Nyag-rong. West of Dar-rtse-mdo the Nyag-chu is joined by the She-chu river (Xianshui). Flowing south in Sichuan, it is known as the Yalung. It joins the Yangtze river in Yunnan.

'BRI-CHU

The 'Bri-chu (Zhi Qu, Tongtian) arises in the lDang-la range. In its upper course it has several branches, the largest of which is known as Muru Ussu or the Chu-dmar. Flowing through eastern Tibet and into China, the 'Bri-chu is also known as the Jinsha Jiang. It becomes the Yangtze as it crosses China. The world's fifth longest river, it empties into the Pacific after a course covering 3430 miles.

RDZA-CHU

The rDza-chu (Za Qu) arises west of bKra-shis-dgon in the mountainous land north of the lDang-la range. It flows through eastern Tibet where the Ngom-chu joins it at Chab-mdo. Flowing south, it reaches Yunnan where it is known as the Lancang Jiang, and entering Southeast Asia it is called the Mekong. It drains into the South China Sea south of Vietnam after a 2600 mile long course.

DNGUL-CHU/NAG-CHU

The dNgul-chu (Nu Jiang) is also known as the Nag-chu in its western course and gives its name to the Nag-chu-kha region. From its source in the lDang-la range, it flows across southeastern Tibet through Southeast Asia, where it is called the Salween. It empties into the Gulf of Martaban south of Burma after a 1730 mile long course.

RGYAL-MO-DNGUL-CHU

The rGyal-mo-dngul-chu (Dajin Jiang) arises in the mountains around gSer-khog south of 'Gu-log, and flows through rGyal-morong. South of Dar-rtse-mdo, where it is called the Dadu He, it turns east and flows south of the mountain Omei Shan to join the Min Jiang. The Min then empties into the Yangtze.

GLANG-PO-CHE-KHA-'BAB

The Glang-po-che-kha-'bab (Sutlej) arises south of Mount Kailāśa on the western edge of the plateau in the Elephant Mountain, a glacier peak in the Himalayas. It flows through Gu-ge, Kulu, Mandi, and Jalandhara to join the Indus river.

RTA-MCHOG-KHA-'BAB (GTSANG-PO)

The rTa-mchog-kha-'bab originates east of Kailāśa on the western edge of the plateau in the Horse Mountain, the Jiema Yangzong glacier in the Himalayas. It flows along the old suture line between the Indian and Asian plates. In its western course it is known as the rTa-mchog-kha-'bab; while crossing central Tibet, it is known as the gTsang-po. It turns sharply in the east and flows south to India where it is called the Brahmaputra, finally draining into the Bay of Bengal after a course of 1800 miles.

RMA-BYA-KHA-'BAB

West of Kailāśa is the Peacock Mountain where the rMa-bya-kha-'bab has its source. Known in India as the Karnali, it flows through sPu-rang and into Nepal, joining the Gogra, which in turn enters the Ganges.

SENG-GE-KHA-'BAB

North of Kailāśa is the Lion Mountain where the Seng-ge-kha-'bab arises. Famous as the Indus in its Indian course, this river flows along the western edge of the Tibetan plateau before making a right angle turn at Gilgit, and crossing Pakistan to the Arabian Sea. It follows a 1980 mile long course.

Although there are at least eight important rivers on the Tibetan plateau, only the four around Ti-se have been given symbolic animal names and treated in detail by Tibetan histories and geographies. Though the rivers in eastern Tibet are much longer than the ones centered around Ti-se in the west, and their courses run all the way across the continent, they are rarely discussed. This may be due to the great renown of the holy mountain and its four rivers, the kha-'bab-chen-po-bzhi, and also to a distinction sometimes made between Tibet and Greater Tibet, Bod and Bod-chen-po. Many similar instances suggest that geographers and historians traditionally treated these areas differently.

As efforts to open the resources of the Roof of the World begin, hopefully the rivers, lakes, and mountains will be preserved, for their natural power and beauty have brought centuries of blessings to the land of Tibet.

MOUNTAINS AND RIVERS IN TIBET

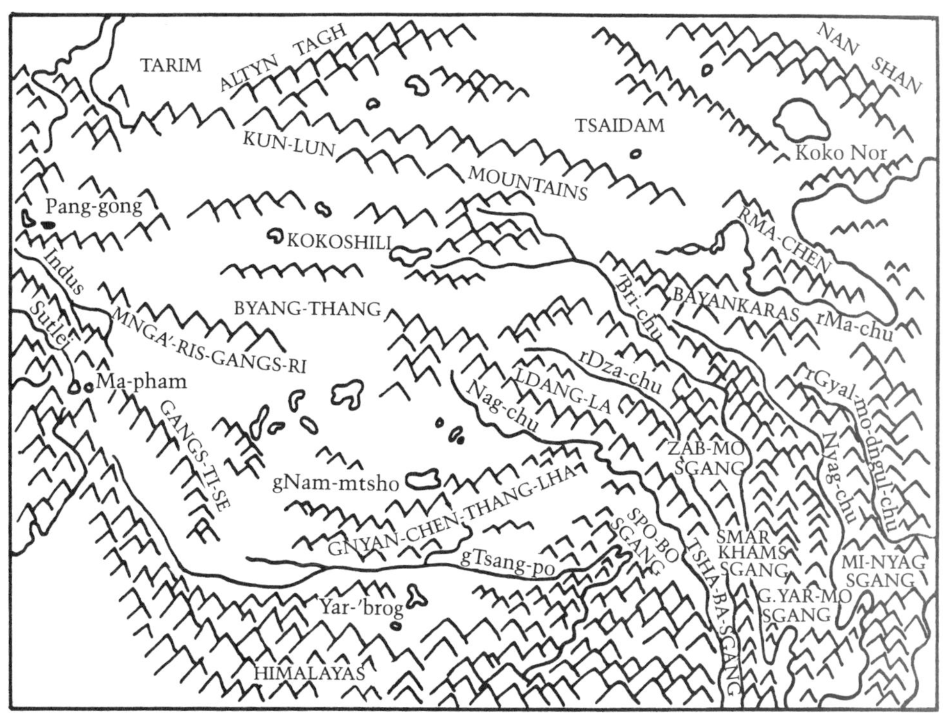

CROSS-SECTION OF THE PLATEAU

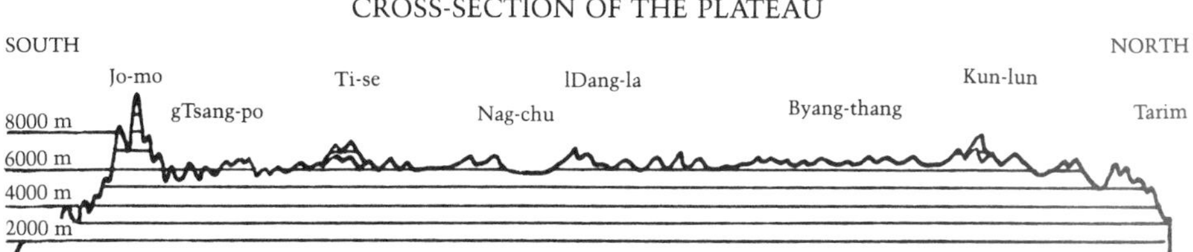

Higher than all of its neighbors, the land of Tibet is truly the Roof of the World. The plateau rises over 15,000 feet above sea level, and the surrounding mountain peaks often climb above 22,000 feet, soaring many thousands of feet above the lowlands of China, Central Asia, and India. Many of the world's highest mountains are found in the ranges around Tibet, especially within the spectacular Himalayas. Some of Asia's longest rivers originate on the Tibetan plateau, which is an important reservoir of water for all of Asia.

FAULTS ACROSS THE PLATEAU

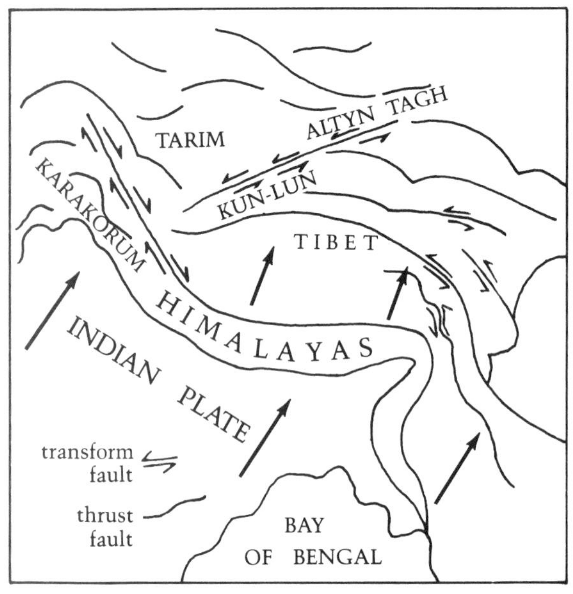

Deep faults across the plateau are thought to coincide with old plate boundaries. Transform faults (strike-slip faults) exist where plates slide past each other, and are found in the Kun-lun mountains along the northern border of the plateau, in the Himalayas and Karakorum ranges, as well as along the 'Bri-chu and the Nyag-chu rivers. Thrust faults occur where plates press against each other, and these exist in many regions of Tibet and in the Himalayas.

EARTHQUAKE ZONES

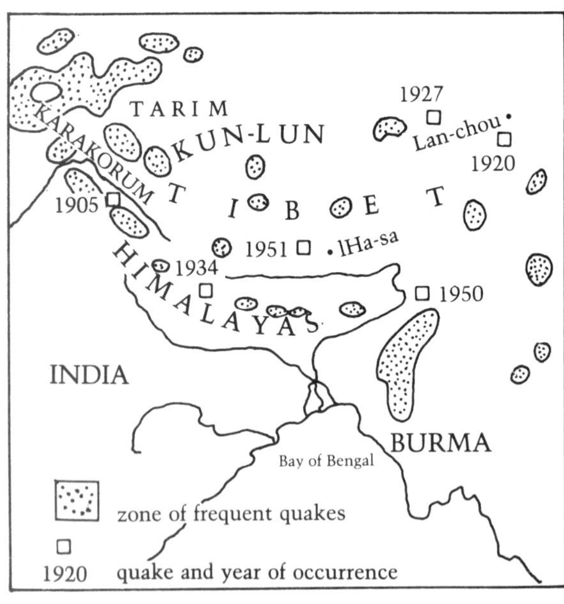

Continuing activity of the crust beneath Tibet is indicated by the frequency of earthquakes of large magnitude. Deeper quakes take place in the mountains around the plateau, while the quakes on the plateau are shallow. In the past century, six large quakes have occurred, rated at least eight on the Mercali Scale, a condition where ground cracks on steep slopes, and towers and chimneys fall. The continuing pressure from the Indian plate is related to quakes across Asia.

A CENTURY OF EARTHQUAKES IN TIBET

1870
'Ba'-thang area

1893
Northwest of Dar-rtse-mdo

1904
Southeast of Brag-mgo

1911
Khyung-lung area

1913
Dro-shod area

1914
Nub-ru area

1915
Western Bayankaras

1915
Southeast of lHa-sa

1918
North of gTsang gNam-ring

1919
Hor-khog area

1919
Southwest of gSer-thal

1920
Southwest of Kokoshili

1921
gNyan-chen-thang-lha area

1923
Brag-mgo area

1923
'Ba'-thang area

1924
Kun-lun range

1924
gNyan-chen-thang-lha area

1926
Western Kun-lun range

1926
Southwest of Kokoshili

1927
East of A-mdo mTsho-nag

1927
Nan Shan range

1930
East of A-mdo mTsho-nag

1932
Northwest of Dar-rste-mdo

1934
lDang-la range

1934
South of Zilling lake

1935
Nag-tshang area

1935
South of rGyal-rtse

1936
Southwest of Lan-chou

1937
Oring-Tsaring lake region

1937
Nub-ru region

1938
Northwest of Koko Nor

1938
sPo-bo area

1938
West of lDang-la range

1940
East of gNam-mtsho

1940
East of gNam-mtsho

1941
East of Dar-rtse-mdo

1944
Ngang-la-ring
3 large quakes

1947
Southwestern Byang-thang

1948
Western Kun-lun range

1948
Li-thang area

1949
Southern 'Gu-log

1950
South of gNas-chen-gangs-ri:
11 quakes

1951
South of Chab-mdo

1951
gNyan-chen-thang-lha area:
4 quakes

1952
East of gNam-mtsho

1952
South Tsaidam

1952
Southern 'Gu-log

1953
Western Byang-thang

1953
North of Dang-ra lake

1955
Western Byang-thang

1955
Dar-rtse-mdo area

1957
West Nag-tshang

1958
West Nag-tshang

1959
lDang-la range

1959
Central Kun-luns

1962
Tsaidam

1966
Southeast Gu-ge

1966
North Byang-thang

1967
Hor-khog area

1971
Oring-Tsaring

1971
South of dGe-rtse:
2 quakes

Quakes greater than Mercali 6

CHAPTER THREE

ANCIENT ENVIRONMENT

The plant and animal life in Tibet appears to have changed gradually over long eras of geologic time in response to changes in climate and environment, following evolutionary patterns that have influenced life all over the world. Scientists describe how a marine environment with tiny shelled animals was replaced by coal swamps and dinosaurs. Slowly, these swamps became warm forested lowland where rhinoceroses and three-toed horses roamed. The high steppeland that today we call Tibet developed during the last Ice Age some three million years ago.

According to the evolutionary description of the earth's history, new life forms have emerged in succeeding ages: Marine algae and fungi were followed by simple land plants and these in turn by complex forests. Primitive marine animals with shells were succeeded by fish, amphibians, reptiles, and mammals.

The theory of evolution, first proposed in the mid-nineteenth century by Charles Darwin, is now well supported by evidence from ongoing research in all the biological sciences. In this view, the present forms of life on the planet are descendants of earlier ancestors in lines that are continuous, though sometimes obscure and uncertain. How and when these lineages spread into Tibet may become clearer as research continues.

Data on ancient animals and plants of Tibet published in 1981 include the identification of fossils and analysis of plant pollen deposits in various layers of the plateau. This information allows scientists to describe the vegetation in Tibet at different times. The following sections summarize these descriptions and connect them to the accepted timeline for the development of animals and plants around the world.

UNDERSEA KINGDOM

The earliest life forms discovered in Tibet so far are simple marine creatures that lived 500 million years ago. They must have had ancestors, and to understand their lineage, we must look at evidence from other parts of the world. One theory suggests that the earliest life began in the seas as simple molecules combined into the more complex ones necessary for proteins, a genetic code, and the organization of cells. Recent evidence dates the first primitive plant cells back at least 4000 million years ago.

As more complex plant cells formed, photosynthesis released oxygen into the environment. This gradually produced a protective atmospheric shield around the earth. Under these new conditions, multicelled animals and plants such as worms and jellyfish arose. Evidence of such early life has been found in Canada, Australia, Newfoundland, and Russia, dating back about 680 million years.

By 500 million years ago, the era of the first marine life from Tibet, seas across the ancient planet were inhabited by snails, shrimplike crustaceans, jellyfish, and clamlike molluscs. The floor of the ocean was covered with corals, sea anemones, sponges, and other stationary animals resembling plants. Especially abundant were trilobites, shelled animals with segmented bodies, as well as brachiopods, also shelled creatures with tentacles, that attached themselves by stalks to the ocean floor. These are the earliest kinds of fossils found on the Tibetan plateau.

In some regions primitive fish had emerged by this time as well. By about 400–300 million years ago, fishlike ancestors had given rise to amphibians that could inhabit dry land as well as the seas. But no evidence of ancient fish or amphibians has as yet been found in Tibet. There are abundant fossils of marine plants, however, which differ from region to region. Many researchers interpret the variations as evidence that cold water covered southern Tibet, while warm-water seas existed over the land destined to become northern Tibet. This fits with the view that southern Tibet was part of Gondwanaland, which was situated far to the south in the cold reaches of the southern hemisphere.

SWAMPLAND OF DINOSAURS

About 245 million years ago as Pangaea was forming, worldwide mass extinctions decreased the number of families of animals by half. Some researchers speculate that a collision with a comet might have been responsible for this mysterious catastrophe, the worst in the history of the world. But even before this time, a new type of animal, the reptile, was developing from the amphibian line. By 230 million years ago, the age of dinosaurs had begun. Other early reptiles began to develop in different directions, giving rise about 200 million years ago to the first primitive, ratlike representatives of the mammals.

It was in this era that northern Tibet was joining Asia. Shallow water covered most of the northern plateau until the north and east began rising out of the sea. Fossils of this period alternate between marine and continental, but land plants had become established. Extensive marshland and swampland developed, where coal deposits were laid down as swamp plants died. The far northeast was drier though still subtropical, and giant horsetail plants and ferns now formed forests in the Kun-lun mountains, across lDang-la, and in parts of Khams.

In this warm and moist climate, tropical plants flourished, and dinosaurs roamed the plateau. Fossils of Changdusaurus, a dinosaur of 150 million years ago, have been found in eastern Tibet in the Chab-mdo area. This prehistoric creature stood some nine feet high and was about twenty feet in length. In southern Tibet Ichthyosaur fossils were also found at Nya-lam. Over thirty feet long, these marine dinosaurs lived about 180 million years ago. Their fossils confirm that far southern Tibet was still under water at that time.

About 100 million years ago the climate was still warm and moist across the plateau, as it was throughout most of the world. The far southern parts of Tibet were mostly under water on the Indian plate, which was moving north. Volcanoes were erupting in the sea between India and Tibet, and Gangs-ti-se and gNyan-chen-thang-lha were forming on the edge of Asia.

In this era neither the plateau nor the mountain ranges were very elevated, and tropical plants common to the northern hemisphere spread across the plateau. Broadleaf forests, which were now evolving, began to replace the more ancient fern and conifer forests just as they had done in other parts of the world. The swampland was becoming forested meadowland.

FORESTLAND OF ANCIENT MAMMALS

The approach of India and southern Tibet toward Asia coincided with the end of the age of dinosaurs. Mammals now began to predominate all over the world. There is not much evidence available yet on early mammals in Tibet, but some scientists speculate that the mountain ranges that separate the plateau from the rest of Asia were not yet high enough to block animal migrations. Thus the plateau's forests could have been inhabited by primitive mammals and ancient birds like those found in other parts of Eurasia in this era, including rodents and squirrel-like primates, dog-sized early horses, and the first hooved animals.

After India collided with Asia about 45 million years ago, the Tibetan plateau was complete, but the environment remained a tropical and subtropical lowland for millions of years. Gradually, the vegetation changed, due to worldwide climatic shifts and climate changes resulting from the rising of the plateau.

By 25 million years ago, the far northeast beyond A-mnyes-rma-chen had become drier, though the climate was still warm. The land was covered with forest and grassland. Hemlock, holly, and ferns, as well as oak forests, grew in the Kun-luns, where the climate was quite humid. Tsaidam was wetter and warmer than today, evidenced by the discovery of fossil ostrich eggs. Forests there were composed of fir, spruce, birch, cedar, beech, and chestnut, in addition to gingkos and magnolias. Across the northern plateau a complex forestland flourished, including oak, hemlock, spruce, fir, pine, cedar, and beech, together with conifers, broadleaf evergreens, holly, ferns, and willows. Central Tibet was much like Tsha-ba-rong today, richly forested with oak, poplar, walnut, willow, cedar, hemlock, sequoia, and cypress. North of the gTsang-po and south of gNyan-chen-thang-lha the land was forested with poplar, birch, hornbeam, oak, and elm, and rhododendrons were plentiful.

About 35–20 million years ago, when forests had reached a maximum worldwide, primates rose to predominance. Primitive monkey fossils have been found in Burma, and true monkeys in Egypt, Africa, and the Americas. Fossils of various ancient apelike creatures that lived between 20 and 7 million years ago have been found in Africa, Europe, Asia Minor, and Egypt, as well as at sites nearer Tibet: in Yunnan in southwest China, in the Siwalik hills in the Punjab, and in Pakistan.

Though no apelike fossils have yet been discovered on the Tibetan plateau, some scientists believe that further investigation may uncover such fossils. Primates may have spread into the plateau's forests, for Tibet was not yet isolated by mountains.

GRASSLAND ANIMALS

Forest declined all over the world some 15 to 10 million years ago as the climate cooled. The evolution of grass from bamboo-type plants, which occurred about 24 million years ago, had prepared the way for new environments.

This shift from forest to grassland was especially marked in Tibet because the plateau was rising; temperatures dropped, and the climate grew even drier. Tropical landscape changed to steppeland with fir, spruce, birch, and hazelnut forests spreading across the northern plateau, mNga'-ris, lDang-la, the Kun-lun mountains, and Tsaidam. The east, however, remained subtropical with oak, maple, mountain ash, pine, and poplar forests, while the Himalayas were quite warm and wet, with oak and cedar forests growing across the slopes. A warm, moist belt existed across Eurasia around the regions where the ancient Tethys Sea had been, and Mediterranean type birds may have lived in Tibet.

The grassland regions across the plateau now supported herds of grazing animals—ancient gazelle, giraffelike animals, and deer. Fossils of Asian rhinoceroses, large cats, hyenas, and rodents have also been found. Primitive three-toed horses were widespread in Asia and Europe at this time and roamed freely across the plateau.

Over several million years, more modern types of animals emerged, and many ancient types became extinct. The little three-toed horse died out, but a related species in North America gave rise to the modern horse. By 2 million years ago, these new horses had spread across the world. In Tibet the horselike kiang was a descendant of the recent arrivals. Modern cattle, sheep, and goats also arose in this era, with the yak developing 400,000 years ago.

THE ICE AGE

A distinct cooling all across the earth began about 3 million years ago. Though there have been many ice ages in the distant past of our planet, this most recent event is what is usually meant by the "Ice Age." It was during this era that humans arose, and culture had its beginnings (see chapter 5).

The Ice Age lasted until 14,000 years ago, but was punctuated by many warmer periods called interglacials that might last 10,000 years or more. As the climate shifted between glacial and interglacial, the vegetation on the Tibetan plateau shifted back and forth between forest and steppeland.

In glacial times, the forests receded from mountain slopes; sometimes sparse spruce and fir forests would survive in sheltered valleys, but much of the landscape was treeless, shrubby grassland and steppe. During warmer periods, the temperature increased rapidly even in the northern areas. As glaciers melted, new lakes formed across the plateau, and older lakes were replenished, while marshland increased. Trees returned to the valleys and lower mountain slopes, and alpine forests of pine, oak, birch, elm, and cedar flourished again.

Since 1960 scientific expeditions have gathered information on glaciers at over twenty locations including Mount Everest. This data can be connected to well-known worldwide climatic changes. About thirty-four cold periods occurred between 3.25 million years ago and 128,000 years ago; the ice grew deeper about 2.4 million years ago. In this early part of the Ice Age, about 1.8 million years ago, glaciers formed on the highest peaks of the Himalayas, but not on the other Tibetan mountain ranges. About 700,000 years ago, the ice pack on the Arctic began to last through the summer months. In this middle period of the Ice Age, the climate in Tibet during the glacial periods was extremely severe, resembling that in polar regions today.

The worldwide decrease in temperature was intensified in Tibet because of the continued rising of the plateau, which was now happening at an accelerated pace. Some investigators believe the plateau had now reached 10,000 feet (3000 meters), and the Himalayas were 15,000 feet (4500 meters). Many mountains in Tibet were covered with glaciers by this time. On the Himalayas, huge glaciers, some of them 24 miles (40 kilometers) across, grew down into the foothills. Most of the slopes of Gangs-ti-se were glaciated as fingers of ice 12–15 miles (20–30 kilometers) long reached down the mountains. Valleys in Khams were filled with extensive glaciers nearly 80 miles (130 kilometers) long, and lDang-la glaciers also increased. As ice came down the mountainsides, it pushed huge amounts of rock and gravel in front of it. When the climate warmed and the glaciers retreated, they left behind these new hills, called moraines.

As the plateau rose higher and higher, the Himalayas became so high they began to block moisture coming from the Indian Ocean. At higher elevations, the air above the plateau was thinner and less able to hold moisture, and solar radiation became more intense. Thus, the climate became less favorable to glaciers, which need moisture to form ice; now snowlines retreated, moving higher up on the mountainsides. From the middle of the last Ice Age, the glaciers on the interior of the plateau were diminished even in glacial periods. Those in the Himalayas and in the southeast, however, still received moisture from monsoons and remained extensive.

Though glaciers across the plateau were decreasing, permafrost became widespread. In this condition, the soil itself freezes and remains frozen for at least a year at a time, except for a thin layer on the surface that may melt during the summer months. The last major advance of the ice began 72,000 years ago. In the Far East, the onset of this cold period commenced about 70,000 years ago. About 58,000 years ago, a mild interlude took place, although this is not yet well documented for Tibet. The climate turned severely cold 28,000 years ago, becoming even harsher than the previous cold period. By 18,000 years ago, the ice reached its maximum.

COLD STEPPELAND ANIMALS

Tibet was greatly changed by the severe Ice Age climate. The increasing height of the plateau harshened the climate to such a degree that even during interglacial periods the weather remained cold and dry, though not so severe as during glacial periods. This in turn altered the animal population. Cold-tolerant birds and animals from the steppes and deserts of northern and central Asia are thought to have moved into Tibet as it became the cold and dry habitat that they were accustomed to. Red deer, musk deer, steppe cats, Persian gazelles, and Mongolian jerboa arrived on the plateau. Horselike kiang from western Asia moved into Tibet, as did wild yak, which were widespread in Asia at this time. Their fossils have been found in northern Asia and even in Alaska, which was connected to Asia during the Ice Age. Because glaciers had locked up water in the form of ice, sea level had fallen, and a land bridge had appeared across the Bering Strait, which allowed animal migration between the continents.

Fossils of other animals, such as Tibetan gazelle, Himalayan marmot, and Tibetan antelope, have been found only on the Tibetan plateau, and not in the rest of Asia. These types are presumed to have developed into their distinctive Tibetan forms on the plateau. When the climate turned harsh, they apparently adapted to it and stayed.

Other animals that could not adapt to the changing environment fled south and east where the warmer forest and meadowland zones remained. The southeast was inhabited by badgers, hyenas, tigers, porcupine, horses, deer, bison, elephants, civets, pandas, takin, macaques, and leopards. It has not yet been determined how many of these animals lived over a wider range of the plateau before the climate became so harsh.

The timing of the development of Tibetan wildlife is still uncertain. Most may be new arrivals during the last Ice Age, but older species seem to exist as well. In most regions of Tibet, scientists generally report both new and old species. Older forms from warmer times sometimes hide out, finding small "tropical zones" in the cold alpine steppe. Aquatic snakes, for example, survive today in hot springs regions. A type of warmth-loving salamander inhabits eastern Tibet, but is found nowhere else on the plateau. Some geese even fly over the Himalayas to breed in Tibet, behavior that might be a holdover from the past when Tibet was much lower and warmer.

A warming trend began all over the world about 14,000 years ago, or 12,000 B.C. Between 8000 B.C. and 5500 B.C., the environment grew more favorable; the variety of plants and animals increased, and the glaciers retreated. The plateau may now have been well over 15,500 feet (4700 meters) in height, and the Himalayas near 20,000 feet (6000 meters).

Between 5500 B.C. and 1000 B.C. the climate became even warmer and more humid than it is now; vegetation was more lush, and glaciers were no greater in extent than they are at present. New lakes formed from melting glaciers, and older lakes became fresher and deeper. Forests of pine and birch grew around many lakes, even in the Himalayas. The far north was still quite cold and not so moist as the south and east, which were more humid and more heavily forested than today. Many lakes and swamps dotted the landscape, and the climate was tropical enough around Chab-mdo for ferns and brake to grow.

About 1000 B.C. a cold period, known as the New Ice Age, set in around the world. Rapidly the climate in Tibet became drier and harsher, similar to the present. Lakes diminished and became saltier, while peat bogs and marshes declined. Forests shifted toward steppe, and steppeland shifted toward desert steppe. The central plateau lost many forests, and those in the northeast diminished. The most drastic change occurred in the south, which had a more lush vegetation during warmer periods. Now the gTsang-po valley became deserty, and Yar-'brog lake, which had been connected to the gTsang-po river, was blocked off with mud and rock. Another especially harsh period occurred in the first century A.D., followed by a third recent cold period between the 17th and 19th centuries. During these times, glaciers advanced, and temperatures dropped worldwide.

The modern environment is the result of many millions of years of climatic changes and innumerable adaptations and migrations of ancient animals and plants. But both the recent Ice Age and the rapid rise of the plateau, which began at the same time, had a profound and lasting effect on Tibet, creating the high steppeland environment we see in Tibet today.

THE THEORY OF EVOLUTION

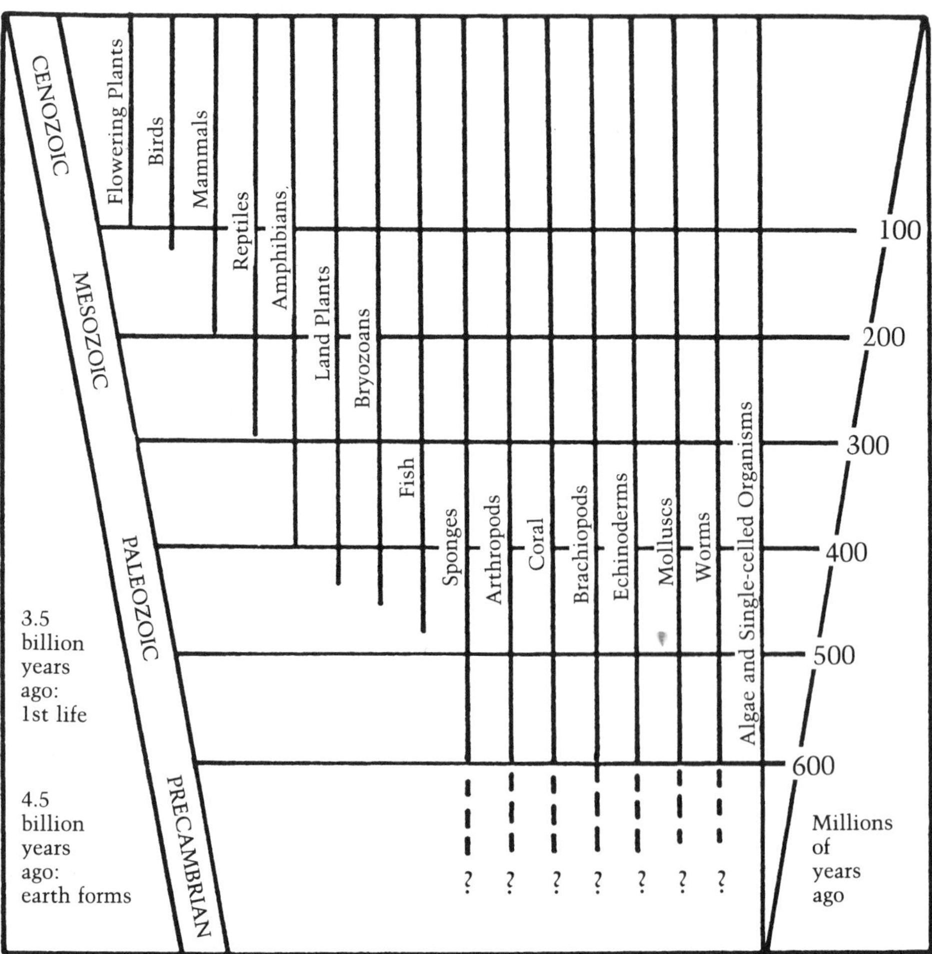

According to the theory of evolution, different life forms arose at different times. Single-celled plants and animals arose first; they gave rise to more complex forms, such as corals, worms, sponges, and seaweed. More advanced animals began to emerge by 450 million years ago. The earliest fish—first animals with a backbone—were followed by amphibians and reptiles. From a branch of the reptile line, mammals arose 200 million years ago. Among the mammals were the primates, which eventually gave rise to humans.

THE RECENT ICE AGE IN ASIA

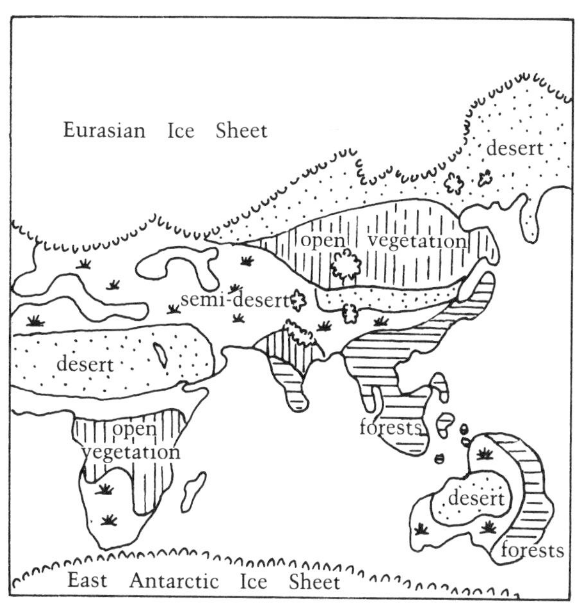

The most recent Ice Age began 3.25 million years ago, in a period when the plateau also began rising rapidly. This combination of events increased the severity of the Ice Age climate in Tibet. Across the world, steppe and desert replaced forests during the glacial periods, but trees and vegetation returned during the warmer interglacials. In Tibet, once the plateau had risen, the cool, dry climate could not support widespread forestland, and steppeland increased.

CAUSES OF THE ICE AGE

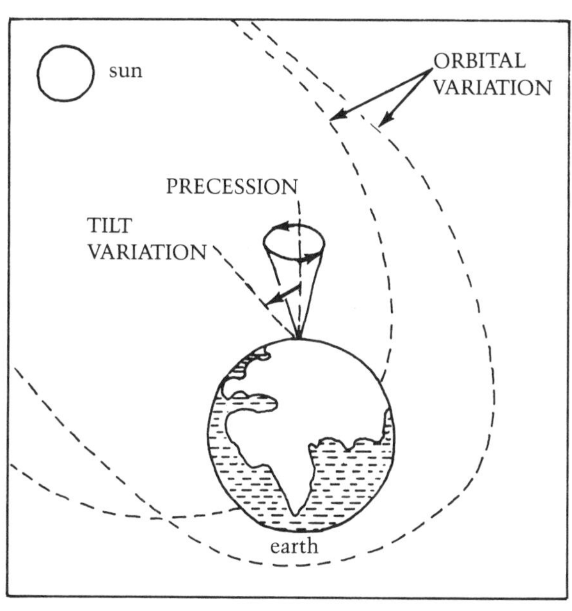

The most recent Ice Age seems to be related to cyclical variations in the orbit of the earth around the sun, to changes in the degree of tilting of the earth's axis, and to changes in gyration of the axis known as precession. Glaciation is favored by cool summers and mild winters, which occur when the orbital is more oval than round, when the axis is less tilted, and when the closest approach to the sun falls in the winter. It is not yet clear whether earlier ice ages fit these cycles.

ANIMAL MIGRATIONS

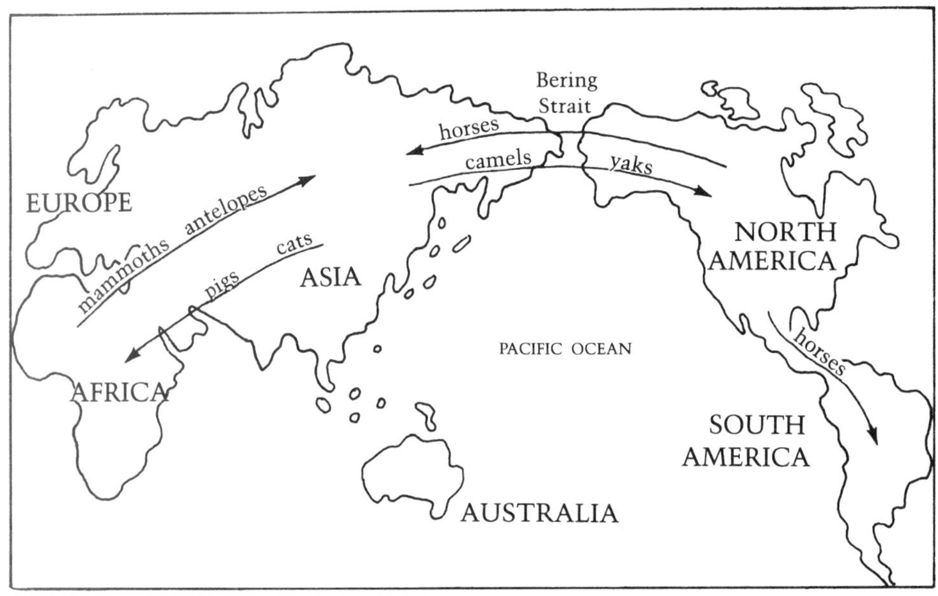

After Africa joined Asia 20 million years ago, animals began traveling between the two continents. Later migrations between Asia and the Americas also took place during the Ice Age when sea level dropped, and a land bridge appeared across the Bering Strait.

EMERGENCE OF ANIMALS

55 mya
Early horse, early rhinos

35 mya
Early cats and dogs, monkeys

19 mya
Early antelopes in Africa; migrate to Asia, give rise to grazing animals by 11 mya

7 mya
Three-toed horse, elephants

4.5 mya
Camels, bears, pigs

3.7 mya
Modern horse in N. America; to Eurasia by 2.5 mya

2 mya
Water buffalo, bison, sheep, modern big cats, wild hogs

650,000 years ago
Wolves

400,000 years ago
Yak

200,000 years ago
Modern cow, goats

ENVIRONMENTAL CHANGES IN ANCIENT TIMES

TIBETAN ENVIRONMENT	WORLDWIDE ENVIRONMENT	TIBETAN ENVIRONMENT	WORLDWIDE ENVIRONMENT
500 mya Simple sea animals 300 mya South Tibet under cold water North Tibet under warm water 200 mya North Tibet rises out of sea: dinosaurs, tropical forests, swampland 100 mya Broadleaf trees replace tropical forests 45 mya India and S. Tibet collide with Asia 25 mya Tsaidam, Kun-luns Byang-thang, east and central Tibet forested 10–5 mya Temperatures drop, climate dries; forests shift to grassland Asian rhinoceros, antelopes, hyenas, deer, gazelle, three-toed horses 3 mya Central Asian animals invade plateau	430–360 mya "Greenhouse" period: warm and humid 290 mya Ice Age 220 mya Supercontinent forms, climate more seasonal; fall in sea level: competition between animals increases 170–70 mya Second "Greenhouse" period 67 mya Sudden extinction of dinosaurs, large land animals 37–25 mya Cold increases N. American forests change from broadleaf evergreen to deciduous 29 mya Second major cooling 9 mya Glaciers form in Alaska 6 mya Glaciers appear in S. America 3.25 mya Recent Ice Age begins	Glacial periods: dry cold steppe, forests decrease Interglacials: warmer, more humid, forests increase 700,000 years ago Climate severe and dry: huge glaciers form 100,000 years ago Climate too dry: glaciers decrease in interior; Himalayan glaciers remain extensive 28,000 years ago Climate turns very harsh 18,000 years ago Maximum of ice 12,000 B.C. Warming trend begins: glaciers retreat Variety of plants and animals increases 8000–1000 B.C. Warmer, more humid than at present: more forests and lakes 1000 B.C. Climate harsher, drier: lakes and forests decrease	700,000 years ago Ice pack in the Arctic lasts through summers 500,000 years ago 29 episodes of glaciation since beginning of Ice Age 128,000 years ago Interglacial 115,000 years ago Glacial 108,000 years ago Interglacial 95,000 years ago Glacial 88,000 years ago Interglacial 72,000 years ago Glacial 58,000 years ago Interglacial 30,000 years ago Glacial 18,000 years ago Maximum of ice 14,000 years ago Interglacial 7800 B.C. Monsoons begin in India 3800 B.C. Sharp cooling 2900 B.C. Warming trend 2500 B.C. North Africa dries

mya = millions of years ago

CHAPTER FOUR

MODERN ENVIRONMENT

Though Tibet is almost entirely highland plateau, the environment varies widely from region to region, and forested mountains, deserty plains, and hilly pastureland can all be found. This great variation depends largely on climate. As climate changes with latitude, fairly regular zones of vegetation are formed at different latitudes around the world. Near the equator, land is usually covered with tropical forests. In middle latitudes north or south of the equator, temperate deciduous forests of beech, maple, oak, chestnut, and hickory are common, as well as grasslands and sometimes desert. In higher latitudes conifer forests of spruce, pine, and fir are predominant. In the polar regions only tundra remains, a zone of permafrost with a thin carpet of grass and some low shrubs.

If Tibet followed this worldwide pattern, the landscape would be forested in the south with deciduous broadleaf trees and perhaps grasslands, while regions more densely forested with conifers would appear toward the north. Across the Tibetan plateau, however, the usual north-south pattern is not regular. The very high altitude, the great dryness, and the complex landforms create local climates that affect the pattern of vegetation. Thus the south is more heavily forested than the north, which is mostly rocky desert; the southeast is much more densely forested than central Tibet, while the northeast is grassland.

Vegetation zones also change with altitude, and high mountains may have a whole series of climates similar to those found in different latitudes. Moving from low elevation to high elevation, mountain vegetation usually shifts in a regular pattern: tropical forest, deciduous forest, grassland, pine forest, and alpine scrub above the tree line, with barren rock at highest altitudes. Mountain ranges in Tibet, however, do not always exhibit the entire sequence of zones. In the southeast, where the climate is relatively moist and warm, mountains show more regular arrangements of forest and meadowland. But often across the plateau mountain vegetation is reduced to only steppe and desert.

A region-by-region description of the Tibetan landscape has been published by the U.S.S.R. Academy of Sciences, and many interesting details on climate and vegetation are also scattered throughout the works of early explorers. Academia Sinica has made data available recently in a special report on the vegetation of Tibet, in the papers from the 1980 symposium, and in a pictorial volume dealing with the vegetation and landscape.

BYANG-THANG

This vast plain stretching across northern Tibet is subjected to strong winds and intense solar radiation. Rainfall is less than four inches (less than 100 millimeters) per year and falls only in the summer, which is brief. Mean July temperature is no more than 50°F. (10°C.). Variations in temperature are great, ranging from 86°F. (30°C.) during the daytime to 5°F. (−15°C.) at night. Winter is long with severe frosts, and temperatures drop down to −31°F. (−35°C.). Snow rarely falls, most of it evaporating because of the dryness. Winds often exceed 45 miles per hour (20 meters per second), creating frequent dust storms.

Rivers in Byang-thang are fed by glaciers that melt in the summer, while in the winter, most rivers freeze to the bottom. Salt lakes and salt swamps are abundant, the thick salt beds often sculpted into unusual shapes by the high winds. Mountain peaks are smooth and rounded, weathered by both winds and frosts. Much of the ground is covered by coarse gravel and pebbly material that has become compacted and shiny.

In the north, rock gravel deserts with sparse vegetation are common. Only shrubby vegetation, such as sage and sedge, grows around lakes with drainage or on mountain slopes with water. The far northwest is high cold desert.

In southern Byang-thang the slightly increased rainfall promotes steppeland pasture with fescue, feathergrass, and quack grass; juniper grows around the southern lakes. Byang-thang is a source of abundant salts of various kinds, such as soda and borax, as well as asbestos, graphite, iron, gypsum, quartz, and gold. Animals adapted to the cold inhabit the plain, including wild yak and wild kiang, also known as wild ass; Tibetan antelope, gazelle, sand foxes, snow leopards, brown bears, wolves, lynx, and rodents are common in the southern regions of Byang-thang. Blue sheep inhabit the rocky regions, especially in the far west. Ravens, tit warblers, snow finches, ground choughs, snowcocks, sand grouse, larks, and desert wheatears are frequently found. Lake Pang-gong in the northwest is the home of brown-headed gulls.

To the north of Byang-thang, the highest peaks of the Kun-lun mountains catch moisture from westerly winds, but the lower and middle elevations are very dry. Landscapes are deserty, and forest and steppe zones are absent in the western Kun-luns. In the eastern Kun-luns, steppe zones appear, as well as small forests of spruce and juniper.

TSAIDAM

Tsaidam (Tshwa'i-'dam) is a basin 46,000 miles square lying 9000–10,000 feet (2700–3000 meters) above sea level in the northeastern corner of the plateau. The landscape is arid with gravel, sand, and clay deserts, and the climate is very harsh, with cold winters. Rainfall is no more than four inches (100 millimeters) per year, and the northwestern region is nearly waterless, a true desert. Thistle and occasionally white willow grow by salt lakes. The southeast is watered by streams from the Kun-lun mountains that flow into salt marshes where reeds, sedge, and thickets of tamarisk grow. The southern borders of Tsaidam support dry pastureland and scrubby vegetation. Tsaidam is rich in coal and oil, as well as copper, tin, gold, and silver. Other resources include borax, potash, bromine, and soda.

SOUTHERN TIBET

This environment is bounded by the ranges of Gangs-ti-se and gNyan-chen-thang-lha in the north and the Himalayas in the south. The climate is warmer and moister than in Byang-thang, but temperatures fluctuate greatly. Long, dry winters can be very cold, down to 7°F. (–14°C.), while summer temperatures reach 60–85°F. (16–30°C.). In some places annual rainfall is 20–40 inches (500–1000 millimeters), most of it falling in summer and early fall. Monsoons reach the south sides of mountains, but the north sides of Gangs-ti-se and of gNyan-chen-thang-lha, as well as small enclosed valleys, are left quite dry.

To the west, the climate is drier and colder. Gar-thog has rainfall as low as that in Byang-thang. Winter temperatures reach 11°F. (–12°C.), and alpine steppes predominate.

The south sides of Gangs-ti-se and gNyan-chen-thang-lha have fairly regular vegetation zones. The regions of alpine steppe are very well developed with many kinds of plants such as juniper, barberry, honeysuckle, and cotoneaster. Wildflowers, clover, sedge, dandelion, and foxtail grow in the meadows. The valleys of the gTsang-po river in its upper and middle course are shrubby steppeland. Shrubs and dwarf willows and poplar can be found across the valleys in gTsang, but true forests begin only in the river valleys east of lHa-sa. To the east, in Kong-po and Dags-po, mountain slopes have abundant coniferous forests, with fir and rhododendron growing at 13,000 feet (4000 meters), hemlock and spruce or larch growing at 10,000 feet (3000 meters), and pine or laurel at 5000 feet (1500 meters).

At the bend of the gTsang-po around the gNam-lcags-bar-ba region, the climate grows even more moist and warm, and forests are complex and extensive, similar to those in the southeast of Khams. Fir, maple, cherry, and hemlock grow on the upper slopes to 13,000 feet (4000 meters); bamboo, evergreen oak, camphor, and magnolia woods grow lower down. At even lower elevations are found tree ferns, banana trees, wild orange, and palm. Tropical forest covers the land below 3000 feet (1000 meters) with myrtle, banyans, orchids, fig trees, and tropical almond trees.

The southern valleys are a source of gold and iron, as well as coal, talc, granite, and magnetite. Animals in the south include foxes, leopards, many rodents, and musk deer. Compared to Byang-thang, the southern valleys are inhabited by a much larger variety of birds, including green snowcocks, bar-headed geese, finches, mountain turtle doves, sun birds, magpies, tree creepers, and snow finches. The more tropical zones to the east merge with the southeastern forest land (see below) and are the home of species found in east and southeast Asia, such as leopards, wild cats, monkeys, bats, and many colorful birds.

NORTHEASTERN TIBET

To the east Tibet is largely mountainous terrain. The climate and temperature vary greatly from area to area, for the ruggedness of the landscape creates small enclosed environments. Rainfall varies from 20 to 40 inches (500–1000 millimeters) per year, increasing toward the south. Thus, the vegetation of the northeast is distinctly different from the southeast.

A-mdo and 'Gu-log in the northeast are drier with great expanses of steppe and meadowland on the mountain slopes. Sedge, rock jasmine, rhubarb, and gentian grow on the steppe, and many medicinal herbs are found among the alpine plants growing above the treeline. Rhododendron, willow, spruce, and juniper grow on lower slopes. The mean winter temperature in Khyer-dgun-mdo is 18°F. (−8°C.); in the far north, at Zi-ling (Xining), temperature in January drops to –16°F. (–27° C.), while mean July temperature is 64°F. (18°C.) with 93°F. (34°C.) the maximum.

At A-mnyes-rma-chen the weather is very changeable, and temperature varies with elevation. July is the hottest month with temperatures reaching 62°F. (17°C.), while January is the coldest month with temperatures of –36°F. (–38°C.). Annual rainfall is about 14 inches (350 millimeters) per year, falling mostly in summer and early fall.

Mineral and metal wealth of the northeast includes gold, copper, lead, iron, coal, chromium, and zinc. Common animals are yak, brown bears, wolves, musk deer, white-lipped deer, woolly hares, marmots, jackals, goats, gazelle, snow leopards, argali sheep, muskrats, and foxes. Goldcrest, quail, rose finches, snowcocks, rose buntings, black-necked cranes, jays, and grouse are especially common birds.

North of A-mnyes-rma-chen are steep, rocky gorges cut by the rMa-chu (Huang Ho), with forests in the valleys of the tributaries. The mountain range itself is covered with glaciers and is surrounded by rich grassland. On the east slopes are forests of pine, spruce, and cypress. To the west the environment begins to merge with that of Byang-thang, while to the south, the tributaries of the rMa-chu flow through broad valleys. At the southeastern bend of the rMa-chu is swampy marshland.

SOUTHEASTERN TIBET

Farther south into Khams, grassland gives way to forested steppe. The treeline in the southeast is at 14,000 feet (4300 meters), and spruce and fir grow at elevations as low as 10,000 feet (3000 meters). Below this are shrub thickets of oak and pine groves. Much of the southeast is coniferous forest with temperate mixed forests and subtropical forests predominating farther east.

Southeastern forestland is very complex, varying greatly from valley to valley in this rugged terrain. Moss, grass, and shrubs grow everywhere. In colder, drier regions toward the north, pine and oak are common. Warmer, wetter regions grow fir and hemlock between 7000–13,000 feet (2000–4000 meters). Below this level are mixed forests of hemlock, tanbark oak, and chestnut, as well as birch and aspen. More than one thousand different kinds of trees can be found, many of them fast-growing species. This, together with a long growing season, makes the southeast a lush zone of complex forestland. Other natural resources of this region are gold, silver, copper, lead, and iron. Animals include white-lipped deer, musk deer, and a few warmth-loving species such as pandas and monkeys. Common birds include grouse, tit warblers, buntings, parakeets, and thrushes.

The far south is quite subtropical, for the ranges in dMar-khams and Mi-nyag receive monsoon winds that bring moisture from the Indian Ocean. In the southern valley of the 'Bri-chu, yew grow at 13,000 feet (4000 meters), spruce, fir, and pine at 10,000 feet (3000 meters), and Yunnan pine and alpine oak below 7000 feet (2000 meters). This zone changes to laurel, magnolia, and other subtropical trees. Animals in this region include more tropical types such as fruit bats, civets, otters, tree frogs, and many monkeys. Pheasants, partridge, squirrels, and deer are also common, and birds such as parakeets, sunbirds, laughing thrushes, minla, magpies, and flowerpeckers are found.

GLACIERS ON THE TIBETAN PLATEAU

The Tibetan plateau and the surrounding mountains are the most extensive area of mountain glaciation in the world today, encompassing 18,190 square miles (46,640 square kilometers) of glaciers. This is half of all the glaciated area of Asia. The presence of glaciers and the level of the snowline in Tibet depend on the amount of precipitation, the atmospheric circulation, and the amount of solar radiation, all of which are interrelated with temperature and elevation.

The high plateau receives large amounts of solar radiation and small amounts of precipitation. Thus, across Byang-thang glaciers are not so extensive as those in the surrounding mountains. Plateau-type glaciers are small ice caps or flat-top glaciers. The high reaches of the western Kun-luns receive large amounts of precipitation from winds blowing in from the west, so glaciers there are very large. The south slopes of the Himalayas, especially in the east, receive more rain from monsoons than the north slopes, and so have more glaciation.

Between 1950 and 1960 glaciers on the Tibetan plateau were reported to be shrinking, as were the glaciers in the European Alps. In the Kun-luns, the Pamirs, the Himalayas, and the Karakorum mountains, glaciers increased in the 1970s, but those in the interior of the plateau did not increase. Of 116 glaciers studied by scientists over the past ten to thirty years, 62 are decreasing and 32 are increasing.

AGRICULTURE ON THE ROOF OF THE WORLD

Tibet has the highest upper limit of agriculture in the world. The climate is especially suited to wheat and barley, and cold-resistant highland barley can be cultivated even at 15,500 feet (4750 meters). The most important agricultural regions are along the Yar-lung and gTsang-po rivers, the Nyag-chu (Nu Jiang), the rDza-chu (Lancang Jiang), and the 'Bri-chu (Jinsha Jiang). The elevation varies in these regions between 9000 and 13,500 feet (2700–4100 meters). Weather is mild and cool in the growing season, and the temperature rises slowly in spring. Rainfall is not extreme and usually is limited to night hours so that there are numerous sunny days. Second harvests are possible in some areas: barley followed by wheat, barley followed by millet, or rice followed by barley.

The high-altitude agriculture is possible because the plateau, at its great height, absorbs large amounts of solar radiation, and the air close to the ground is heated by radiation from the land. The temperature on the plateau is higher, especially during the growing season, than on steep mountains in other parts of the world at the same elevation and latitude.

Rice and maize can be grown only at lower elevations; thus rice cultivation in Tibet is possible in the deep valleys in the southeast and around the huge bend in the gTsang-po where it turns south toward India. Rice crops are much more common in neighboring regions such as the south slopes of the Himalayas and in Yunnan and southeast Asia.

In addition to cereals, peas, and potatoes, which grow in many areas, certain regions are especially suited to crops that are not widespread. Apples, walnuts, and pears grow in the central gTsang-po valleys, while tea, grapes, tangerines, bananas, and oranges grow in the southeast. Jujube and apricots are plentiful in mNga'-ris in the west. At lower elevations in many regions, cabbage, tomatoes, cauliflower, onion, garlic, celery, radishes, turnips, and strawberries can be grown. Recently, additional kinds of crops have been cultivated experimentally.

VAST NATURAL PASTURELAND

The grasslands of Tibet are a great natural resource. Tibetan grass is especially rich in nutrients and makes excellent fodder for domestic herds as well as wild animals. Good grazing land exists even at 16,000 feet (4900 meters) in the east, at 17,000 feet (5200 meters) in the central plateau, and at 18,000 feet (5500 meters) in the west. Between the Himalayas and Gangs-ti-se can be found mountain grassland consisting of temperate zone grasses and drought-resistant species. Alpine meadows cover great expanses of the plateau, and make the best grazing land because sedge and sagebrush survive even when there are no frost-free periods. Plateau grassland in the far north consists of needlegrass and can also be used for grazing land. Marshy meadows near small pools and springs, especially in warmer regions, grow sagebrush and other abundant grasses that are available at the end of winter when other pastures are depleted.

Across these pasture zones graze yak, sheep, goats, and horses. Sheep predominate in the west, and horses are especially common in A-mdo and Khams, while southern Byang-thang and southern A-mdo are habitats for yak, as well as goats and sheep. Cattle and donkeys are raised especially in the south.

The yak, which is a special breed of ox, is a particularly valuable animal, for its hair can be used to make tent cloth, and its skin for clothing; its dung can be burned for fuel in treeless regions. The sturdy physique of the yak makes it an excellent pack animal as well. A special crossbreed of yak and cow, known as the dzo, has been developed in Tibet. Yak and sheep provide meat, milk, and cheese, which make up a large part of the Tibetan diet. Often soup of boiled meat and barley is the main meal of the day for inhabitants of the pasturelands.

During summer, animals can be moved into higher elevations, and nomads often change location each month. But during the winter, the highlands become too cold, and animals must graze at lower elevations. Camps set up in the foothills last the whole winter. Expeditions are then made to markets early in winter where farmers and herdsmen trade animal products such as cheese, butter, and meat for farm produce, especially barley and dried fruits.

The forestland, fertile valleys, and pastures that support the animal life and inhabitants of Tibet are among the most precious resources of the plateau. This vast pristine natural environment is closely tied to a whole style of life and has provided the culture with the raw materials for its arts, sciences, and crafts. The metals and lumber have been used for centuries in architecture and statuary, while the minerals and precious metals play a role in both medicine and art. New uses of these resources are being found, and Tibet may benefit from both the traditional and modern ways of life, as long as the natural environment is carefully preserved.

WILDLIFE IN TIBET

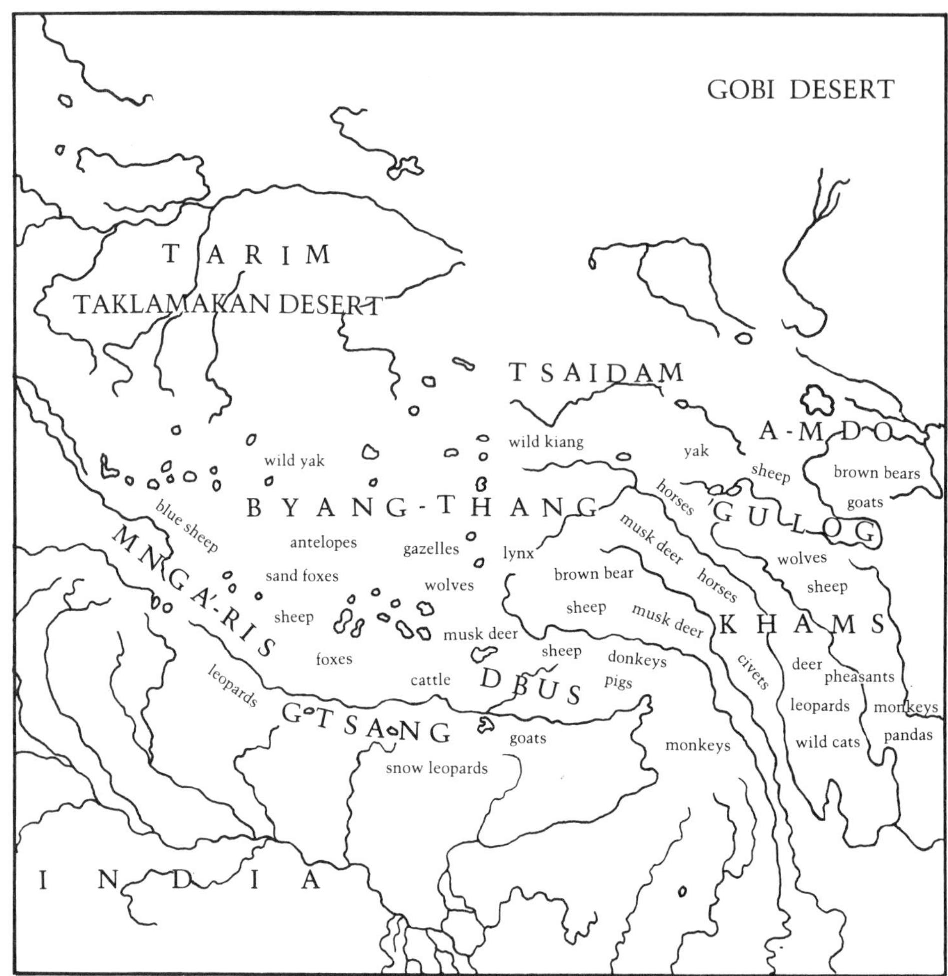

The Tibetan plateau exhibits a variety of habitats for wildlife. The southern part of Byang-thang, though dry, supports wild grazing herds. A-mdo and 'Gu-log are cool and dry, but have extensive grasslands and some wooded areas. Farther south in Khams the climate becomes warmer and more humid, and animals are more like those in southern Asia. dBus and gTsang are not so humid, but warmer and wetter than Byang-thang. The environment at the eastern end of the gTsang-po merges with Khams. mNga'-ris is steppe and pastureland.

DIMENSIONS OF THE TIBETAN PLATEAU

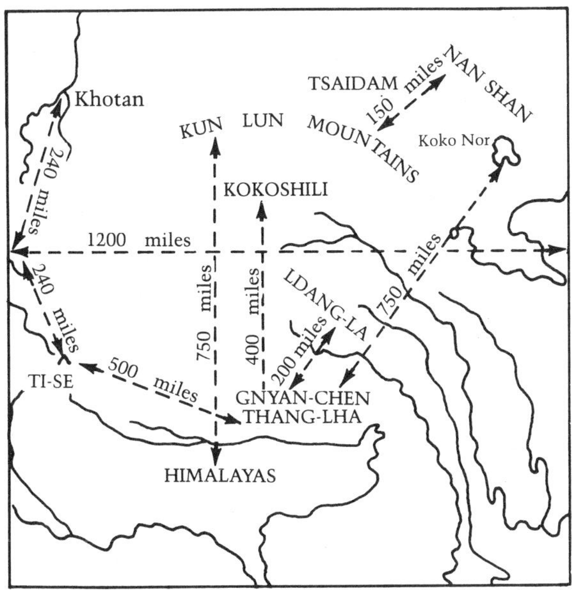

The Tibetan plateau, the highest land mass on earth, extends between the Kun-lun mountains in the north and the Himalayas in the south. To the west it is bounded by the Karakorum range while to the east the high mountains extend past rGyal-mo-rong up to the Cheng-du plain before the elevation drops. Over one million square miles in area, the plateau is just slightly smaller than India, twice as large as Iran, and four times as large as the state of Texas.

TRAVEL DISTANCES ACROSS TIBET

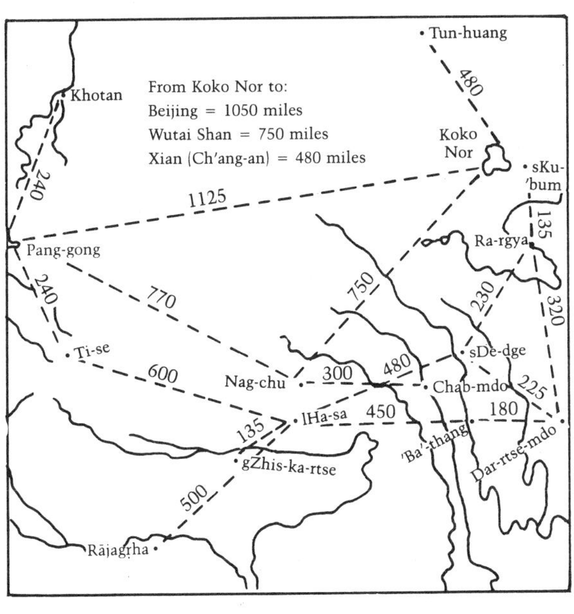

Many hundreds of miles separate the important cultural centers of Tibet, and long-distance travel has always been necessary. In earlier centuries, caravans regularly made their way from the capital at lHa-sa across the lDang-la mountains north toward Koko Nor, a journey that might require sixty to ninety days. The trip from Ti-se to lHa-sa would also take about that long. To reach Beijing in far northeastern China from lHa-sa could require eight months.

RAINFALL ZONES IN TIBET

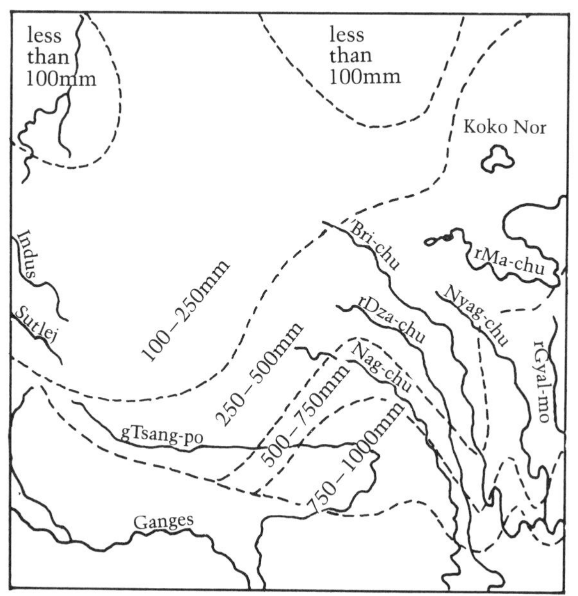

The amount of rainfall varies greatly across Tibet, being higher in the southeast where the monsoons penetrate, decreasing steadily toward the northwest. Vegetation patterns are closely related to rainfall. Lush forests in the valleys of the southeast give way to pastureland and dry steppe across the northern plains. The climate shifts again toward cold desert in the far northwestern borderlands. The far northeast also turns frigid and dry in the Tsaidam basin.

VEGETATION ZONES IN TIBET

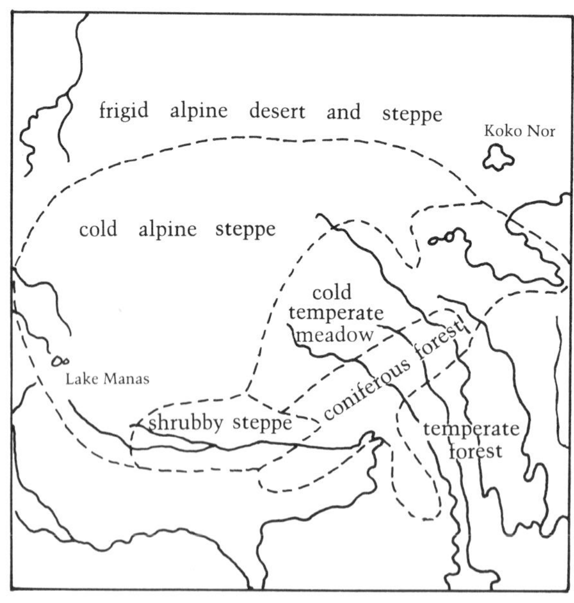

Unusual climates and vegetation patterns of Tibet do not follow standard worldwide latitudinal zoning. The shifts are due partly to large variations in elevation and rainfall, and partly to the complexity of the landforms. High and rugged mountain chains enclose many valleys creating local environments. Location of mountains blocks the flow of air currents that would bring more moisture from the south into the center of the plateau. Altitude adds to the dryness and coolness.

PLANTS AND ANIMALS IN TIBET

Across the Tibetan plateau have been found over 530 species of birds, over 190 species of mammals, more than 40 species of reptiles and 30 species of amphibians, in addition to 2300 species of insects. The following lists, though far from complete, offer a sampling of the wealth of animal and plant life in Tibet.

BIRDS

Alashan Redstart
Alpine Accentor
Austen's Barwing
Azure-winged Magpie
Bar-headed Goose
Bar-winged Wren Babbler
Bar-throated Minla
Bearded Vulture
Beautiful Sibia
Beavan's Bullfinch
Black-chinned Yuhina
Black-faced Laughingthrush
Black-faced Warbler
Black-headed Greenfinch
Black-headed Jay
Black-headed Sibia
Black-headed Tree Babbler
Black-necked Crane
Blanford's Snowfinch
Blood Pheasant
Blue-eared Pheasant
Blue-fronted Redstart
Blue-throated Sunbird
Blyth's Leaf Warbler
Brandt's Rosy Finch
Bronzed Drongo
Brown Accentor
Brown Bullfinch
Brown Creeper
Brown Ground Chough
Brown-headed Gull
Brown-headed Tit Babbler
Brown Parrotbill
Buff-bellied Flowerpecker
Chestnut Thrush
Chestnut-crowned Laughingthrush
Chestnut-headed Tesia
Chestnut-headed Tit Babbler
Chinese Babax
Chinese Copper Pheasant
Chinese Hazel Grouse
Chinese Nuthatch
Cinnamon Sparrow
Cinnamon-breasted Tit
Coal Tit
Collared Grosbeak
Common Hill Partridge
Common Pheasant
Common Quail
Coral-billed Scimitar Babbler
Crested Tit Warbler
Crimson Tragopan
Crimson-browed Finch
Dark-breasted Rosefinch
Dark-rumped Rosefinch
Daurian Jackdaw
Daurian Redstart
Derby's Parakeet
Desert Finch
Desert Lesser Whitethroat
Desert Wheatear
Dusky Warbler
Elliot's Laughingthrush

Eurasian Nuthatch
Eurasian Snowfinch
Eurasian Tree Sparrow
Eurasian Woodcock
Fire-capped Tit
Fire-tailed Myzornis
Fire-tailed Sunbird
Firethroat
Fulvetta
Fulvous Parrotbill
Fulvous-fronted Crowtit
Giant Babax
Giant Laughingthrush
Gold-naped Finch
Golden-headed Tree Babbler
Gould's Sunbird
Gray Bushchat
Gray-crested Tit
Gray-faced Warbler
Gray-sided Laughingthrush
Great Parrotbill
Great Tit
Greater Yellownape
Green Magpie
Green Shrike Babbler
Green-backed Tit
Golden-spectacled Warbler
Golden-throated Barbet
Gould's Shortwing
Grandala
Ground Chough
Guldenstadt's Redstart
Hill Prinia
Himalayan Brown Crowtit
Himalayan Edible-nest Swiftlet
Himalayan Greenfinch
Himalayan Monal Pheasant
Himalayan Rubythroat
Himalayan Tit
Hoary Barwing
Hodgson's Rosy Finch
Horned Lark
Hume's Ground Jay
Ibisbill
Indian Blue Robin
Indian Tree Pipit
Kalij Pheasant
Kessler's Thrush
Koklas Pheasant
Koslow's Babax
Koslow's Bunting
Lady Amherst's Pheasant
Large-billed Bushwarbler
Long-billed Calandra Lark
Long-tailed Tailor Bird
Long-tailed Tit
Minla
Maroon-backed Accentor
Mongolian Plover
Mongolian Finch
Mongolian Lark
Mountain Tit Babbler
Nepal Cutia
Nepal Treecreeper
Northern Three-toed Woodpecker
Orange Crowtit
Orange-barred Warbler
Orange-bellied Leafbird
Pere David's Laughingthrush
Pheasant Grouse
Pied Cuckoo
Pink-rumped Rosefinch
Plain Flowerpecker
Plain-backed Thrush
Prince Henry's Laughingthrush
Przewalski's Rosefinch
Pygmy Blue Flycatcher
Pygmy Wren Babbler
Raven
Red-bellied Tragopan
Red-billed Chough
Red-billed Leiothrix

Red-breasted Hill Partridge
Red-breasted Rosefinch
Red-headed Bullfinch
Red-headed Crowtit
Red-headed Laughingthrush
Red-headed Tree Babbler
Red-necked Snowfinch
Red-tailed Minla
Red-winged Shrike Babbler
Red-whiskered Bulbul
River Chat
Robin Accentor
Roborovski's Rosefinch
Rock Partridge
Rock Sparrow
Rose Bunting
Rosefinch
Rosy Finch
Rosy Pipit
Ruddy Shelduck
Rufuous-breasted Accentor
Rufous-breasted Bush Robin
Rufous-capped Babbler
Rufous-cheeked Hornbill
Rufous-necked Hornbill
Rufous-necked Laughingthrush
Rufous-necked Scimitar Babbler
Rufous-vented Tit
Rufous-vented Yuhina
Rufous-winged Fulvetta
Rusty-cheeked Scimitar Babbler
Scaly Laughingthrush
Scaly-breasted Wren Babbler
Scarlet Minivet
Sclater's Monal Pheasant
Serin
Sibia
Silver-eared Leiothrix
Sinai Rosefinch
Smoky Leaf Warbler
Snow Finch
Snow Pigeon
Spangled Drongo
Spectacled Parrotbill
Spotted Munia
Spotted-breasted Scimitar Babbler
Spotted-winged Grosbeak
Spotwinged Rosefinch
Stonechat
Streaked Great Rosefinch
Streaked Laughingthrush
Streaked Spiderhunter
Streak-throated Fulvetta
Stripe-necked Fulvetta
Stripe-throated Yuhina
Stoliczka's Tit Warbler
Szechwan Jay
Temminck's Tragopan
Tibetan Eared Pheasant
Tibetan Partridge
Tibetan Sandgrouse
Tibetan Serin
Tibetan Snowcock
Tibetan Snowfinch
Tickell's Leaf Warbler
Treepie
Variegated Laughingthrush
Yellow-bellied Fantail
Yellow-breasted Greenfinch
Yellow-browed Tit
Yellow-hooded Wagtail
Yellow-naped Yuhina
Yellow-streaked Warbler
Yellow-throated Fulvetta
Wedgetailed Green Pigeon
Western Tragopan
White Wagtail
White-bellied Redstart
White-browed Bush Robin
White-browed Rosefinch

White-browed Shortwing
White-browed Tit Babbler
White-cheeked Nuthatch
White-crested Laughingthrush
White-eyed Tit Babbler
White-rumped Snowfinch
White-spotted Laughingthrush
White-tailed Nuthatch
White-throated Laughingthrush
White-throated Redstart
White-throated Warbler
White-winged Grosbeak
Willowtit
Wood Snipe

ANIMALS

Argali Sheep
Asian Golden Cat
Asiatic Black Bear
Asiatic Chipmunk
Asiatic Shrew
Asiatic Shrew-mole
Asiatic Striped Squirrel
Asiatic Water Shrew
Barking Deer (Muntjac)
Bharal
Birch Mouse
Black-lipped Pika
Blue Sheep
Brook's Pika
Brown Bat
Brown Bear
Burrowing Vole
Central Asian Gazelle
Chinese Desert Cat
Clouded Leopard
Desert Hamster
Dhole
Dzo (Yak/Cow cross)
Field Mouse
Flying Fox
Giant Flying Squirrel
Giant Panda
Goral
Gray Wolf
Ground Squirrel
Hare
High Mountain Vole
Himalayan Black Bear
Himalayan Tahr
Hog Badger
Horse
Horseshoe Bat
Indian Civet
Irrawaddy Squirrel
Jird
Jackal
Jumping Mouse
Jungle Cat
Langur
Leopard Cat
Lesser Oriental Civet
Lesser Panda
Long-eared Bat
Long-eared Jerboa
Long-tailed Leaf Monkey
Long-tailed Mole
Long-tailed Shrew
Long-nosed Tree Squirrel
Lynx
Marmot
Masked Civet
Midday Gerbil
Mole-rat
Mongolian Jerboa
Muntjac
Murree Vole
Musk Deer
Narrow-skulled Pika
Old World Badger

Old World Porcupine
Pallas' Cat
Pere David's Vole
Persian Gazelle
Pika
Pine Mouse
Pipistrelle
Pygmy Hog
Rat
Rat-like Hamster
Red Deer
Red Fox
Red-bellied Tree Squirrel
Red-cheeked Squirrel
Rhesus Macaque
River Otter
Rock Squirrel
Roe Deer
Root Vole
Sand Fox
Serow
Short-nosed Fruit Bat
Shrew Mouse
Sikkim Mouse
Small-clawed Otter
Snow Leopard
Steppe Cat
Steppe Polecat
Stone Marten
Striped Hyena
Takin
Tibetan Antelope
Tibetan Gazelle
Tibetan Kiang (Wild Ass)
Tibetan Pika
Tri-colored Squirrel
Tubed-nose Bat
Tufted Deer
Urial
Weasel
Webbed-footed Water Shrew
White-lipped Deer
White-maned Serow
White-toothed Shrew
Wild Hog
Wild Yak ('Brong)
Wooly Flying Squirrel
Wooly Hare
Yak
Yellow-bellied Weasel
Yellow-throated Marten

PLANTS

Abelia Twinflower
Acanthopanax
Agapetis mannu
Alkali Grass
Alpine Rhododendron
Altingia
Arrowgrass
Arrowwood
Asia Bell
Asiatic Poppy
Asparagus
Bajiao Banana
Bamboo
Banyan
Barberry
Bedstraw
Bell Flower
Bent Grass
Bergenia
Betelnut Palm
Bird Cherry
Blueberry
Bluegrass
Bracken
Bramble
Breadfruit

Bridal Wreath
Brook Anemone
Brooklime
Buttercup
Butterfly Bush
Camphor
Campion
Cardamom
Chaulmoogra
Chinacane Bamboo
Chinese Cane
Chinese Poppy
Chinquapin
Cinquefoil
Cobresia
Corydalis
Cow Parsnip
Crape Myrtle
Crazyweed
Crementhodium
Clearweed
Dandelion
Delphinium
Dendrobium chrysanthum
Dragonhead
Duabanga
Edelweiss
Entada
Evening Primrose
Evergreen Oak
False Lupine
False Tamarisk
Ferns and Tree Ferns
Fescue Grass
Fig
Figwort
Fir
Fountain Grass
Four-petal Rose
Fritillary
Gentian
Gerbera Daisy
Ginseng
Globeflower
Glory Bower
Golden Chestnut
Goldenray
Gooseberry
Goosefoot
Gourd Vines
Grape
Greenbriar
Groundsel
Heartleaf Viburnum
Heliotrope
Hemlock
Himalayan Birch
Himalayan Cassiope
Himalayan Fir
Himalayan Hemlock
Himalayan Mayapple
Himalayan Pine
Holly
Honeysuckle
Iris
Japanese White Birch
Joint Fir
Knotweed
Kumaon Iris
Ladyfern
Larkspur
Latticeleaf
Leptodermis
Likiang Spruce
Lily-of-the-field
Mahogany
Maple
Mare's Tail
Meadow Rue
Milk Vetch
Mint
Mountain Ash
Moss
Mushroom

Myrtle
Needlegrass
Nepeta
Oak
Onion
Orchid
Padritree
Pea Shrub
Pearl Everlasting
Peppergrass
Pepperroot
Phoebe Nanmu
Pincushion Flower
Pink
Plum
Point Vetch
Pondweed
Poplar
Primrose
Purple Bauhinia
Raspberry
Rhaphidophora
Rhododendron
Rhubarb
Ricegrass
Rock Cress
Rock Jasmine
Rockfoil
Rose
Rosebriar
Rosewood
Sabina
Sagebrush
Sal Tree
Sandwort
Saussurea (Snow Lily)
Saxifrage
Sedge
Shag-spine Shrub
Sibiraea
Sikkim Larch
Sikkim Spruce
Solomon's Feathers
Solomon's Seal
Sophora
Spice Bush
Spiraea
Spruce
Spurge
Starwort
Stonecrop
Sycamore
Tanbark Oak
Taron Maple
Terminalia
Thermopsis
Thoroughwax
Threeawn Grass
Tibetan Draba
Tibetan Figwort
Tibetan Rhubarb
Tigerflower, Toothwort
Tree Rhododendron
Trisetum
Tropical Almond
Usnea (lichen)
Violet
Virgin's Bower
Waldheimia globifera
Wallflower
Water Millfoil
Weigela
Whorlflower
Willow
Wood Betony
Woodfern
Woodland Strawberry
Wood Sorrel
Wild Black Pepper
Wild Lemon
Wild Orange
Wilson Maple

TIBETAN AND NEIGHBORING ENVIRONMENTS

TIBET BYANG-THANG	TIBET EASTERN TIBET	NEIGHBORS NORTH INDIA	NEIGHBORS EAST CHINA
Arid and cold; temperature fluctuates Intense solar radiation Strong winds and severe frosts Salt flats and salt lakes Gravel deserts with sparse plants Pasture in south: wild yak, antelope, sand fox, snow leopard, yak, sheep, goats TSAIDAM Barren basin, waterless SOUTHERN TIBET Warmer, moister; temperature fluctuates Rain in summer and fall; long dry winters Alpine steppe, meadowland, with forests on lower slopes Fox, leopard, musk deer, cattle, donkeys, yak, sheep, goats	Mountainous and rugged Northeast: drier, colder Southeast: moister, warmer Rain in summer and fall Alpine steppe, meadowland, pastureland, with forests on lower slopes Musk deer, deer, yak, sheep, hare, marmots, horses, panda, monkeys, civets, otters AGRICULTURE Up to 15,000 feet: cold resistant barley 13,500–9000 feet: wheat and barley Lower elevations: rice and maize PASTURE West: grazing land at 18,000 feet Central: grazing land at 17,000 feet East: grazing land at 16,000 feet	Arable land along Ganges river valley Tropical forest Frost in winter Very hot summers Rainfall during summer monsoon as ocean air drawn in over South Asia Winter monsoon dry winds over North Asia blow outward across India northeast to southwest CENTRAL ASIA Taklamakan desert in Tarim basin Dzungarian basin less than 1000 feet elevation Turfan depression 500 feet below sea level Dry, deserty steppe and desert ringed by mountains Hot, dry summers; very cold, dry winters with high winds Rainfall similar to Byang-thang, but temperatures warmer in winter, warmer in summer	Sichuan basin lowland under 3000 feet Hot, humid summers; warm, humid winters Rain all year round, but especially in summer Subtropical forest and temperate mountain forest Rice, mixed crops FAR EAST No dry season, very hot summers Tea and rice NORTH CHINA Huang Ho river valley region Hot, dry summers; very cold, dry cold, dry winters Low rainfall mostly in summer; varies from drought to flood Farming in river valleys: winter wheat and millet; 8 month growing season Grassland and steppe; some grazing land; few forests

PART TWO

THE PEOPLE

PERSPECTIVE

THE STUDY OF TIBETAN HISTORY

For centuries, Tibetan culture was transmitted orally, and many customs have been preserved into modern times without having been written down — ceremonies for marriage, birth, and death, rituals connected with the seasons and with special locations, methods for interpreting omens and symbols, procedures for creating binding agreements, guidelines for the design of buildings, statues, paintings, and ritual objects.

The reciting and retaining of oral accounts was the province of individuals who were trained to know their material accurately and to pass it on from generation to generation. These trusted experts transmitted the cultural heritage not only by preserving rituals and ceremonies, but also by recounting the stories that explained how a ceremony was first used, how a village came by its unusual name, or how an ancient king had begun a certain custom. Songs celebrated the deeds of great heroes and told of the exploits of famous ancestors. Thus the history of the people was woven into poetry and preserved.

Though modern civilizations rely almost exclusively on written records, oral traditions flourished throughout the ancient world. Fundamental works such as the *Iliad,* the *Odyssey,* and the Old Testament are thought to have been preserved orally during many centuries before being written down.

The traditions of most European peoples remained oral throughout medieval history. For example, the first written document in archaic French was an oath sworn in 842 A.D. The Anglo-Saxon tribes in Britain began to put their oral traditions into writing in the sixth century when Christian missionaries brought them the Latin alphabet; even so, the first historical annals written in English were not composed until the days of King Alfred about 900 A.D.

Tibet had begun documenting its own history in writing three centuries earlier when the seventh century Tibetan king Srong-btsan-sgam-po commissioned his chief minister Thon-mi Saṁbhoṭa to devise an alphabet and grammar based on the linguistic systems of India. Though oral transmission remained important, some accounts, especially governmental records and works of the king, began to be committed to writing.

A number of chronicles and documents from the era of the seventh to ninth century Dharma Kings were thus available to later Tibetan historians. These authors often noted that their information came from old records, and sometimes they included in their histories long citations from rare documents or material now lost. Valuable records, both written and oral, were partly preserved in the most ancient texts, such as the Maṇi bka'-'bum, the sBa-bzhed, and the bKa'-thang-sde-lnga, all of which survive today. Inscriptions from this era also offer useful records.

A treasure house of Tibetan material from the era of the Dharma Kings was discovered in 1907 at Tun-huang, a Central Asian city that came under Tibetan rule in the eighth century. In an annex of a cave temple, walled up since the eleventh century, were found documents piled some ten feet high, apparently removed from a library or chancellery. Amidst many copies of Buddhist texts, banners, and documents in Sanskrit, Khotanese, Chinese, Sogdian, and Uighur, were found Tibetan court records, chronicles about the early kings, and many miscellaneous records and texts in Tibetan.

The Tun-huang materials offer us documentation contemporary with the expansion of the Tibetan Empire between the seventh and ninth centuries, and also record some of the old traditions concerning the founding of the royal dynasty in the Yar-lung valley many centuries earlier. Though variations exist among certain kinds of accounts, and dates do not always agree exactly, the Tun-huang records support the main outlines of ancient history as reported by later authors, and are considered reliable and trustworthy—as far as they go.

It is obvious that the Tun-huang materials are an incomplete record. Many documents appear to be notes, rough drafts, or copy exercises rather than completed, systematic works. The Tun-huang chronicles, which have come to us in a somewhat systematic form, appear to have been patched together from different accounts that may not have all been written by the same author. Well-known stories about the kings, attested to by other old records, are missing from this compilation.

Even the year-by-year record known as the Tun-huang annals is less than a full cultural record. It is a tersely written summary concerned largely with military and administrative events, such as the installment of a new minister, the taking of a census, or the dispatch of the army to some far-off battlefield. The inner workings of the king's court are reduced to notations such as the king's residence at the opening of the year, the birth of a prince, or a conference with a foreign ambassador.

The very existence of this material in a frontier outpost, some seven hundred miles north of the capital, indicates that extensive records were being kept by the government. Documents from central Tibet, where the Tibetan royal court was located, would obviously be more complete, especially regarding the cultural development of the kingdom.

Fortunately, other ancient records are available. The oldest Tibetan records explain that documents belonging to the kings' inner court were often hidden away for safekeeping. Some of these have been found, such as the records concealed in the pillars of a lHa-sa temple. Srong-btsan-sgam-po's records discovered in the twelfth century show that he himself had copies of his own writings hidden in various locations and ordered that important Bon documents be buried as well.

Detailed information about important individuals and events can also be found in monastic records, while numerous local chronicles recorded the histories of different districts in Tibet. Genealogies kept by many families offer important chronologies, and folk stories contain valuable information about the history and customs of the people. The Bon tradition also has preserved abundant material related to early times in both oral and written form, though these records have scarcely been studied and evaluated in depth even by most Tibetan historians.

Recently, Western scholars have also begun to study ancient Tibetan history and already have made valuable contributions. Some of the more recent Tibetan historical accounts have been translated, and documents from Tun-huang and other Central Asian sites have been published, together with provisional translations and discussions.

But Westerners wanting to study the Tibetan tradition are faced with substantial difficulties. Academic programs in Tibetan language, culture, religions, and history have only recently been established at a few universities. The understanding of the Tibetan cultural heritage that they offer is naturally going to be second-hand. Yet many Tibetan texts assume a great deal of familiarity with the topics involved, and material is easily misunderstood out of context.

Additional problems arise with documents written before the time of King Ral-pa-can in the mid-ninth century. This material often employs archaic spellings as well as poetic conventions and vocabulary that even experienced Tibetan scholars may have difficulty recognizing. So much of this difficult material from Tun-huang remains to be carefully studied that preliminary translations may turn out to be misleading. Western and Tibetan scholars need to consult together to develop accurate translations as more material is studied.

It is also difficult for foreign scholars to be sure their sources are reliable and their information complete. Since relatively few Tibetan accounts are available in the West, Westerners may be gaining a biased view on certain topics. If a subject of particular interest to a Western specialist is not dealt with in detail in one Tibetan history, does this mean that no fuller records exist? In some cases, written records may indeed be scanty. But topics not controversial to a Tibetan audience are often treated very cursorily by Tibetan historians. Or perhaps the author simply was not interested in the topic. A fuller discussion might exist in another text, which the researcher may not have access to or know about.

Keeping in mind these limitations, Western scholarship, with its analytic approach, can raise important questions and be very helpful in clearing up inconsistencies. But healthy skepticism can turn unduly critical. Modern scholars tend to distrust the oral tradition and criticize Tibetan historians when they rely upon it. If old records exist, the style in which they are written or the way they have been compiled leads some Westerners to distrust them or even dismiss them as fabrications of later authors.

Another stumbling block is the heroic style and content of some ancient material. If a king is said to possess great magical powers, or his accomplishments are of mythic proportions, his very existence becomes suspect. Important records are then dismissed on the grounds that they must be folktales. Such "reasonable theories" depend partly on an inability to allow that a few truly remarkable and inspired men could have accomplished so much so quickly.

For example, some Western scholars doubt that Srong-btsan-sgam-po was actually a great Buddhist teacher and patron of the Dharma. According to this view, he was actually a strong supporter of older traditions, who became curious about Buddhism only in the last years of his life after his marriage to a Buddhist Chinese princess. Such theorists have difficulty finding evidence of the king's devotion and understanding of Buddhism because they have dismissed the works he authored as historically untrustworthy. Inscriptions carved on pillars in the time of the Dharma Kings also proclaim him as the founder of the Dharma in

Tibet. But all such authentic records are explained away, while speculation and guesswork are relied upon instead.

Although Tibetan historians may not take such theories seriously, less informed people might. Responsibility for developing an accurate understanding of ancient Tibetan history, in the light of the questions raised by Western scholars, rests largely on the shoulders of Tibetan historians.

In addition to correcting misunderstandings, Tibetan historians are faced with the task of studying the issues left unexplored by the Tibetan historical tradition. Beginning in the ninth and tenth centuries, Buddhist historians began to focus almost exclusively on Buddhist topics. Even though they must have had access to many more documents than remain to us, they often presented only brief summaries of older traditions, leaving inconsistencies unresolved and crucial issues not fully explained. Today, so many hundreds of years later, solid research into ancient periods is thus made much more difficult.

It is also uncertain whether the interpretations of these later authors accurately reflect the original meaning of accounts that must have been many centuries old by the time Buddhism was introduced. If their interests and concerns as Buddhist scholars did not encompass the old folk traditions, they may not have studied them thoroughly. This lack of interest in preserving the ancient history of the people created gaps and contradictions in historical accounts, which leaves them open to criticism.

On the other hand, modern Tibetan historians tend to concentrate exclusively on recent history and political events. These topics are important to document and discuss, but an historical vision limited to politics and power struggles will not preserve and enrich Tibetan culture. In the meantime, the culture and history of the people themselves await thorough investigation.

To develop a comprehensive understanding of Tibetan history, exploration of certain neglected topics seems essential. For now, we have been unable to spend the time and effort to do justice to these topics, which would require extensive research in the most difficult sources. The oldest literature and the Tun-huang documents, for example, are gold mines of information, but to approach them with confidence would require years of preparation. We can nevertheless raise these issues in the hope that Tibetan historians will make increasing efforts to explore them.

An especially difficult area of research is the origin of the first inhabitants. Connections to other life forms such as animals, nature spirits, or other humans is presented differently in different accounts that need to be considered and interpreted carefully.

Buddhist historians do not provide much detail on how these ancestors developed into the four great tribes of Tibet, or what became of each group. Is it possible to reconstruct the histories of each tribe from other sources, determine their native territories, and trace their interactions with each other? If the gradual branching of the tribes and their locations at different points in time could be worked out, many questions could be answered about the relation of the Tibetan tribes to other cultural or ethnic groups on the Tibetan plateau or in neighboring lands.

We do know that many local rulers emerged, holding power in different regions, and that a king finally united them. But even here the events in the process of unification are not well documented. The connection between the kings' lineage and the lineages of the people is also worthy of investigation. The claim that the kings did not intermarry with the Tibetan population for many generations needs to be examined.

One recent effort to cover some of these topics, though not all of them, is Khetsun Sangpo's latest volume, rNa-ba'i-bdud-rtsi, which had just been published as this book went to press. He has assembled a variety of historical materials, which look very useful. He presents the early history of Tibet based on his translations of important Tun-huang materials and on discussions of later historians. His chronology for the kings in the seventh to ninth centuries may not prove completely accurate when all the reliable sources, including the dated portions of the Tun-huang annals and the T'ang annals, are compared. Nevertheless, his contribution to Tibetan historical studies will certainly encourage both scholars and nonspecialists in the future.

As scholars develop a clearer picture of ancient times, it may be possible to integrate historical evidence with archeological evidence. By carefully examining accounts about ancient times, Tibetan historians could suggest the most likely locations where archeologists might make interesting discoveries, and help them develop a framework for interpreting any finds. If physical evidence could be compared with whatever written records are available, a vivid portrait of ancient times might emerge.

In other parts of the world, archeological researchers have uncovered extensive evidence of old cultures in the form of human fossils, tools, pottery, ornaments, and dwellings that have been carefully studied, classified, and organized into coherent sequences. Surprising finds have sometimes allowed so-called legends to take their proper place in history. For example, the ancient Chinese dynasty of Shang (c. 1500 B.C.) was thought to be a literary fiction until the discovery in the 1920s and 1930s of Bronze Age artifacts and divination bones inscribed with the names of the Shang kings. The actual ruins of the mythical city of Troy have also been unearthed in the last century.

Today we can only place the meager archeological evidence and the textual information side by side without being able to draw any but the simplest conclusions. Eventually as the past of Tibet becomes more widely studied, and scholars combine their knowledge and resources, they will surely be able to offer a more comprehensive and accurate picture of the earliest era of Tibetan history. Perhaps one day archeological investigations will join with careful textual research to bring some of Tibet's ancient heroes to life for modern people.

WESTERNERS ENTERING TIBET

1603
Portuguese merchant d'Almeida in Ladakh

1624–1632
Jesuit missionaries in Gu-ge and gTsang: d'Andrade, Cabral and Cacella

1661
Missionaries Grueber and d'Orville travel through China to lHa-sa

1707–1745
Capuchins in lHa-sa: d'Ascoli and de Tours della Penna and da Fano; Tibetan dictionary compiled

1715–1717
Jesuit cartographers make survey of Tibet

1716–1733
Jesuits in lHa-sa: Desideri (1716)

1774–75
George Bogle from British East India Company sent by British Governor-General of Bengal, visits gZhis-ka-rtse

1783–1792
Samuel Turner sent on trade mission to Tibet by British Governor-General of Bengal

1811
Thomas Manning visits lHa-sa

1823
Csoma de Körös in Zangs-dkar

1846
Lazarist fathers Huc and Gabet travel through A-mdo to lHa-sa

1846
Beginning of attempts to explore Tibetan highlands by British, French, Russian, Scandinavian, and American parties

1879–80
Przhevalski from Russia to A-mdo, Khams, Tsaidam, Hami, and Central Asia

1885–86
Carey from British India to Tsaidam, Tun-huang, Kucha, and Hami

1889–90
Bonvalot from France to Lob Nor, A-mdo, Khams

1890
Dutreuil de Rhins from France to Tengri Nor, accompanied by Grenard

1895
Hedin from Scandinavia to Byang-thang

1896
Wellby and Malcolm to Koko Nor, Kun-luns, Leh, and Śrīnagar

1900/1906
Hedin to western Tibet and central Tibet

1900
Koslov from Russia to Koko Nor, Tsaidam, southern A-mdo

1906–07
M.A. Stein and Paul Pelliot obtain manuscripts from Tun-huang

THE WRITING OF HISTORY

c. 750 B.C.
Hesiod defines 5 classical ages of human history

c. 600 B.C.
Later Chou dynasty annals written

c. 484 – c. 420 B.C.
Herodotus, Greek author of 1st known western historical work

c. 460 – c. 400 B.C.
Thucydides, Greek author, writes history of Peloponnesian war

145 – 79 B.C.
Ssu-ma Ch'ien, Chinese historian, collates old records into one chronicle

64 B.C. – 17 A.D.
Livy, Roman historian, writes *History of Rome*

56 – 115 A.D.
Tacitus, Roman historian, writes *Annals* and *Histories*

538 – 594 A.D.
Bishop Gregory of Tours: author of history of the Franks, written in Latin

6th century A.D.
History of the Goths written in Latin vulgate by priest Jordanes

7th century A.D.
Srong-btsan-sgam-po and Thon-mi Saṁbhoṭa create alphabet for Tibetan; 1st histories, court records, Maṇi bka'-'bum

712 A.D.
Record of Ancient Matters: 1st written Japanese material and earliest surviving work on Japanese history

720 A.D.
Chronicles of Japan

731 A.D.
Bede writes *Ecclesiastical History of the English People*

848 – 901 A.D.
Alfred the Great establishes *Anglo-Saxon Chronicle* as register of events

915 A.D.
Arab historian al-Tābarī writes a universal history

c. 1010 A.D.
Book of Kings by Firdawsi, 1st Persian national epic

c. 1050 A.D.
Ssu-ma Kuang writes history of China from 500 B.C. to 1000 A.D.

1135 A.D.
Geoffrey of Monmouth writes *History of the Kingdom of Britain* in Latin

12th century A.D.
Kalhaṇa chronicles Kashmir's history

Mahāvaṁsa: history of Buddhism in Ceylon

1207 A.D.
Villehardouin writes history of Crusades; 1st history as prose genre in French

1332 – 1406 A.D.
Greatest Arab historian Ibn Khaldun writes *Universal History*; outlines philosophy of history

17th century A.D.
Jesuit missionaries from China report great age of eastern civilizations

Great age of Egyptian civilization recognized

Dispute in Europe over antiquity of human civilization

17th century
Over 70 chronologies for human culture developed in Europe

Histories of various cultures disagreeing with Bible are labeled fabulous by some European scholars

1697
Pierre Bayle writes *Historical and Critical Dictionary*; Western historical writing desecularized

18th century A.D.
Mabillon and other Benedictine monks critically examine authenticity of historical documents

1725 A.D.
Giambattista Vico writes *New Science*; history proclaimed best way to understand human nature

19th century A.D.
Study of history envisioned as rigorous science based on textual criticism in Germany, France, Britain, U.S.

20th century A.D.
'Annales' historians emphasize historical social science

CHAPTER FIVE

THE EARLIEST INHABITANTS

The identity of the earliest inhabitants of the Tibetan plateau is still a mystery. Evidence of human habitation during the Old Stone Age has been found at sites in Tibet, but the original homeland of these people and their connection to later inhabitants are as yet unknown. Just as in many parts of the world, the relationship between prehistoric peoples and more recent populations is not well defined.

If we think in terms of modern evolutionary theory, any prehistoric inhabitants of Tibet would ultimately have to trace back to primitive human and even prehuman ancestors. Scientific theories linking the lineages of humans, monkeys, and apes are especially interesting to the student of Tibetan history, for traditional Tibetan accounts describe how the Tibetan people descended from ancestors who had arisen from the descendants of monkeys (see chapter 6).

The general outline of human emergence from apelike ancestors, based on fossils found around the world, is well accepted in the West, and descriptions of the various types of early man can now be found in general reference works. But the relationships among these types are not yet established, leaving some of the most interesting questions unanswered.

Evolutionary theory generally states that primitive types gave rise to more advanced types, which finally gave rise to modern humans. But how and where each new type first emerged is uncertain. Did all ancient populations gradually grow more advanced, or did changes occur in just a few groups, which then flourished at the expense of older populations? Apparently distinct populations were sometimes living at the same time in different parts of the world—did these groups arise from a common ancestor? And which group gave rise to the next most advanced form? Anthropologists have worked out numerous possible family trees, trying to link the known fossils into coherent sequences, without reaching much agreement.

LINK BETWEEN APES AND HUMANS

Anthropologists believe the line that gave rise to monkeys, apes, and humans arose from some uncertain common ancestor about 70 million years ago. This line eventually split several times, with the forerunners of apes separating from the forerunners of the Old World monkeys sometime between 40 and 20 million years ago. Another split is thought to have occurred between 15 and 5 million years ago when the first prehumans emerged from the ape line.

The earliest prehuman creature was believed for many years to be *Ramapithecus,* an apelike animal who lived between 14 and 7 million years ago. But most scientists now believe that *Ramapithecus* is too old to be in the direct line of human ancestors. The great similarity in blood chemistry and genetics between humans and modern apes seems to suggest that the human and ape lines separated only 8 to 4 million years ago. Dispute about this crucial era in human evolution is difficult to resolve because of the lack of fossil evidence.

The oldest creature agreed to be a direct forerunner of human beings is a 4 million-year-old *Australopithecus* from Africa, who is believed to have walked on two legs and lived on the ground instead of in the trees. Several types of later *Australopithecus* existed in Africa until about 2 to 1.5 million years ago, but which type gave rise to the true humans is not certain.

ORIGINAL HOMELAND

Based on these findings and other evidence, most researchers now consider Africa the homeland of the human species. But it seems possible that remains of prehumans might await discovery in other parts of the world. Evidence of ancient ape-men has been found in China, not far from the Tibetan plateau, but it is not yet clear whether the dating and classification of these fossils will correlate with standards accepted by scientists around the world.

Recently Lawrence Swan from America and Jia Lanpo from the Academia Sinica have independently proposed that the Tibetan plateau might be the location of man's emergence. Though prehuman fossils have not yet been found in Tibet, Swan points out that historically animal species expand from north to south; this suggests a northern homeland instead of Africa. He also notes that Tibet was a suitable environment for apelike human forerunners 5 to 15 million years ago when the human line was emerging, and since then has changed more than any other place on earth. Further investigations in Tibet, he feels, may improve the understanding of human evolution.

THE EARLIEST HUMANS

Most researchers agree that the oldest primitive but true human discovered as yet anywhere in the world is *Homo habilis,* a fossil human from Africa about 2 million years old. The oldest fossils of the more advanced and clearly human *Homo erectus,* who lived between 1.5 million and 300,000 years ago, have been found in Java and Indonesia as well as Africa. In Africa *Homo erectus* coexisted briefly with *Homo habilis* and with the last of the Australopithecine ape-men.

Later fossils of *Homo erectus* have been discovered at several sites in Africa, China, India, Southeast Asia, and in Europe. By 500,000 years ago, this primitive man was capable of planning and cooperating in big game hunts and was adaptable enough to live in a variety of climates. He had learned how to use fire and could construct dwellings and make simple clothes from animal furs. To make the tools he needed, he chipped stone into axes and knives.

Between 300,000 and 30,000 years ago, more modern humans with larger brains — sometimes called archaic *Homo sapiens* — were living in Europe, Asia, and Africa. Another group usually classified as early *Homo sapiens* is the Neanderthal peoples. Though physically stockier and more powerfully built than modern humans, the Neanderthals actually had larger brains than we do. They lived in Europe and Asia between 120,000 and 40,000 years ago, and for 80,000 years Neanderthal culture was predominant across Eurasia. They are thought to have communicated well with each other, though anthropologists speculate that their speech may not have been so rapid as modern man's. They shared religious beliefs and rituals, evidenced by elaborate graves. How the Neanderthal populations and other archaic human beings were related and whether they interbred are unresolved issues in anthropology today.

MODERN HUMANS

Where and how modern men first arose is still uncertain. Evidence from genetic mapping, which compares the traits of aboriginal peoples, suggests an original homeland in western Asia, somewhere between the Caspian Sea and the Indian Ocean. Such genetic analysis seems to indicate that all the various groups of modern human beings did not evolve independently over hundreds of thousands of years from different ancient lineages. According to this view, all people today have descended from common ancestors who lived no more than about 50,000 years ago. Where this line of modern man began is debated — perhaps from a branch of Neanderthals, from some population of archaic *Homo sapiens*, or from some other as yet unknown ancient lineage tracing back to *Homo erectus*. As more fossils are discovered, a clearer picture may emerge.

Research in Europe indicates that by 35,000 years ago, the Neanderthals had disappeared, while fossil evidence from around the world shows that modern humans were soon widespread. They lived in Europe by 35,000 years ago, Africa and Siberia by 32,000 years ago. By 19,000 years ago, they had reached North America, having apparently crossed the Bering Strait.

Anthropologists describe how human culture took great steps forward in the hands of more advanced humans, who were physically indistinguishable from modern people. Finer stone tools were developed, and the bow and arrow were invented. Skins and furs were sewn together to make tents and clothing and even boats. The world's first known artistic tradition began at this time. In Europe walls of caves, as well as small artifacts, were painted and engraved, and small figurines were carved.

EVIDENCE FROM THE TIBETAN PLATEAU

Stone tools found at Kokoshili in the north and Ding-ri in the south provide the earliest evidence of human habitation in Tibet. Those from Ding-ri are flake tools made with a stone hammer, while those from Kokoshili are more ancient crude pebble choppers. The tools are clearly from the Old Stone Age, the most ancient phase of human culture. This era extends from almost 2 million years ago until the end of the last Ice Age 10,000 years ago, and includes the tools made by all primitive humans as well as early modern man.

Publications on archeology from the Academia Sinica in 1980 have placed the Tibetan tools from Kokoshili in the Old Stone Age, without any specific estimate of date. Those from Ding-ri are put either in the middle part of the Old Stone Age, contemporary with the Neanderthals in Europe, or in the late Old Stone Age when modern man emerged 40,000 years ago.

Where did the Stone Age inhabitants of Tibet come from? Did they develop from earlier, as yet undiscovered, primitive humans on the plateau? Did they migrate from some other homeland? Until additional research is done throughout Tibet, the age and origins of the people who made these tools thousands of years ago will remain uncertain.

Evidence of early modern man has just recently been discovered on the Tibetan plateau in the Tsaidam basin west of Koko Nor lake. Archeological investigations reported in 1985 mention stone scrapers, knives, drills, and axes, together with tools made of bone and horn dated to 33,000 B.C. These artifacts were found at Xiaochaili lake in the center of Tsaidam where freshwater shells more than 38,000 years old were also discovered. Though Tsaidam is a dry, salty desert today, in ancient times it had abundant vegetation and animals, numerous freshwater lakes, and heavy rainfall—a suitable environment for human beings.

TRACES OF ANCIENT COMMUNITIES

Knowing the Tibetan plateau was inhabited during the Old Stone Age, we might expect the descendants of these early peoples to be living in Tibet in later times. Along the eastern end of the gTsang-po river and in the regions of Kong-po and sPo-bo, traces of primitive communities have been found, including both cave dwellings and "nest" dwellings. Archeologists at work in Tibet have reported how these "nests" were built around a central wooden pole that supported a roof of bamboo and wood, while mats of straw plastered with mud formed the walls. Caves that appear to have been inhabited in ancient times have also been noted by Western explorers at Luk, lHa-rtse, Yar-'brog, Yar-lung, and in Byang-thang.

The era when the "nests" and caves were inhabited has not yet been determined. In the future, intensive field work, especially along the gTsang-po and its tributaries in central Tibet, may allow a more detailed description of these early settlements.

These regions in central Tibet are traditionally associated with the rise of Tibetan civilization. The first kings made their court in Yar-lung near the gTsang-po river, and the very first Tibetan tribes are said to have lived in the same region. In the eighth century, King Khri-srong-lde-btsan came upon unusually colored clays and measures of green rice as he made the foundation for bSam-yas monastery not far north of Yar-lung. Ancient-looking objects in bizarre and interesting shapes often were uncovered when foundations were laid for monasteries, but many of these finds have long since been lost. If historians and archeologists could assemble any of these ancient artifacts that remain, their study might offer helpful clues to the past.

Other regions of Tibet have yielded evidence of prehistoric peoples who were clearly early modern humans. Small, specially shaped stones have been discovered recently near Zilling lake in Byang-thang, at Nya-lam, Chab-mdo, and Nag-chu. Finely chipped pieces of stone such as these have been reported by archeologists in many parts of the world. The small stone blades were affixed to shafts to make tools such as harpoons or spears, suggesting that these sites may have been hunters' camps. The exact age of this Tibetan material is uncertain. Archeologists associate this style of tool, known as microlithic, with Middle Stone Age hunting and gathering cultures, which generally arose as the Ice Age ended shortly before 10,000 B.C.

FARMING CULTURES

As the Ice Age drew to a close, a great advance in human civilization occurred in several parts of the world—the development of farming. Recently published studies now place the planting of carefully selected crops as early as 8000–9000 B.C. at Jericho in Palestine, spreading from the Near East into Europe by 6500 B.C. Another center of early farming developed in the lower Indus valley in modern Pakistan, and another in China by 5500 B.C., perhaps spreading from the farming culture that had begun in Thailand as early as 8000 B.C.

Farming practices began to change the very structure of societies. Anthropological research around the world has shown that together with the permanent settlements of farmers came the rise of specialized social classes, more varied styles of life, and more refined technologies than roving bands of hunters possessed. In addition to planting and harvesting grain and vegetables, the farmers tamed and eventually domesticated animals.

Many decades of research on several continents show that in some parts of the world, the older hunting cultures were gradually replaced by the new farming cultures. As fields were depleted, the farmers opened up more and more forestland to make new fields. In Tibet forests of game and fertile farmland would often have been located together in the river valleys, but the competition between hunters and farmers may not have been intense. Huge expanses of grassland, too dry for intensive agriculture, would have been available to hunters. Wildlife was plentiful in this natural pastureland, especially wild grazing animals such as kiang, yak, sheep, and antelope. According to Tibetan chronicles (see chapter 10), farming was not widely practiced until the time of the ninth Tibetan king about two thousand years ago when irrigation techniques were developed. So it seems likely that hunting societies endured for thousands of years.

Some early tribes on the plateau did settle down to become farmers, for evidence of farming villages has been discovered in several locations in and around Tibet. Near Chab-mdo three dwellings found in 1978 have been dated to 3500 B.C. The houses were built with animal shelter on the ground floor and living space on the second floor, a design still in use in Tibet today. At this same site, stone artifacts, pottery, and bone needles were found. Archeologists believe these inhabitants lived mainly by hunting, but also practiced farming.

A site discovered in 1958 at Nying-khri, north of the gTsang-po river in the Kong-po and Nyang river region, yielded human bones that also belong to this early farming era. Later investigations in 1975 uncovered fifteen stone artifacts and over one hundred pottery shards.

FAR NORTHEASTERN SITES

Other ancient villages have been found in the far northeastern edges of the plateau, at sites known as Pan-shan and Ma-chia-yao along the Tao river in the Co-ne region, while not far east of Koko Nor lake other village remains have been found at the site Ma-ch'ang. Specialists in Asian anthropology and archeology have dated Ma-chia-yao to about 3000 B.C., Pan-shan to about 2500 B.C., and Ma-ch'ang to about 2000 B.C. These farmers cultivated millet and wheat, and kept domesticated animals. Villages were built with a regular plan, but settlements moved around. Well-made baskets, tools, jewelry, and distinctively decorated pottery have been found.

Another ancient village, dated to 1850 B.C., has been located along the Klu-chu (Tao river) at the site Ch'i-chia-ping not far south of Lanzhou. A number of villages of similar culture have also been found along the bSang-chu river, southwest of Tsong-kha and east of Rong-po. The inhabitants were farmers who also kept domestic animals; their houses were built on terraces along the river. Weaving sticks, needles, beads, spindles, and copper objects, as well as decorated ceramic pottery, have been found. Bones of cattle, sheep, and pigs were used in divination, and animal sacrifices apparently were made. Evidence of domesticated cattle and sheep, together with the remains of wool products and bronze tools have been found in Tsaidam at Ta-li t'a-li-ha on the Nomuhung river at a site dated to 1800 B.C.

Archeologists have connected some of these sites, such as Ma-ch'ang, Ma-chia-yao, and Pan-shan, with farming cultures that had developed across northern and western China. Some researchers also see similarities between Pan-shan pottery and Ukrainian pottery of this era, while Ch'i-chia-ping pottery is said to resemble Siberian pottery. Whether these sites in the far northeast are closely related to early Tibetan culture may become clearer as archeological exploration of the Tibetan plateau continues. With knowledge of the prehistoric era of the Tibetan tribes so incomplete, it is not yet known how well the ancient tribal territories correlate with modern political or ethnic boundaries.

WORLDWIDE ADVANCES

The settled way of life gave rise to true cities in a few locales where geographical conditions allowed the development of irrigation. Only on floodplains of large rivers could early farming techniques produce enough crops to support large populations. About the same time that the villages were built in Chab-mdo, the Sumerian civilization was emerging in the Fertile Crescent region of the Tigris and Euphrates rivers, now inhabited by the Moslem Arabs of Iraq. Writing, advanced irrigation techniques, and large-scale architecture were developed, and cities of as many as 10,000 people were established.

Egypt, too, by 3000 B.C. was developing a literate, urban culture, centered around the Nile river. Harappan civilization in western India emerged by about 2700 B.C. at various sites along the Indus river. The first high civilization in Europe was the Minoan culture of Crete, which emerged about the same time as Egyptian civilization. In the far east cities were well established along the Huang Ho with the rise of the Shang civilization in China by 1500 B.C., about the time bronze came into use for tools and ornaments. Thus literate urban societies were arising in several parts of the world in the same era.

NOMADIC WAY OF LIFE

The pastoral, nomadic way of life in Asia seems to have developed only after farming arose in the Near East. Anthropologists have determined that sheep and cattle were first domesticated by settled farmers in the Near East, perhaps as early as 7000–8000 B.C. Bands of hunters in steppe regions unsuited to agriculture began to develop a mobile, pastoral culture that was flourishing by 3000 B.C. This way of life centered around the seasonal migration patterns of animal herds.

Just north of Tibet grassy steppeland stretches along the north side of the Tarim basin to Mongolia. To the west this natural pastureland extends along the north side of the Iranian plateau, past the Caspian and Black Seas across southern Russia, all the way to Hungary. Herds grazing on this steppe are sources not only of meat and dairy products, but also fiber for textiles, and hides, dung for fuel in treeless regions, transportation for people and goods, and power for hauling.

The nomadic lifestyle required more highly organized social structures than simple hunting bands. Herds had to be protected against wild animals and other nomadic tribes, and whole communities had to be moved at the proper times. Anthropologists usually describe early nomadic society as warlike, in contrast to the more peaceable, settled farmers. When men from the steppes learned to harness horses to chariots, about 1700 B.C., they overran Europe, western Asia, and India. Not until after 900 B.C. was the technique of riding horseback perfected, but this skill gave the warriors of the steppes a tremendous advantage over their settled neighbors. Indo-European tribes ruled the steppes until 100 A.D., followed by Turkish and Mongolian nomads.

Though the relationship between the farming, hunting, and nomadic ways of life in ancient Tibet is not yet clear, it seems quite possible that early Tibetan civilization was not unlike the steppe culture to the north. Large stretches of the Tibetan highlands are open grassland, and Tibet has historically had a large nomadic population. It has not been determined when domesticated horses were first used in Tibet. One recent research report indicates that the yak was domesticated in Tibet about 2500 B.C., while evidence of domesticated sheep and goats has been found in several farming sites on the plateau. Specialists in prehistoric Asian art note that the style of many ancient ornaments and rock carvings found in Tibet suggests close links to the early Central Asian steppe culture. Carvings of animals, men on horseback, and warriors, similar to Central Asian artifacts, have been found in Khams, in Ladakh, and in gTsang. Many ancient but undated objects in this style have been discovered—buttons, buckles, bells, pendants, and beads, as well as bronze figurines of animals such as lions, monkeys, and various birds. Future archeological investigations may be able to clarify the relationship between ancient Tibetan and Central Asian cultures.

MEGALITHIC SITES

Other prehistoric sites in Tibet include a number of locations where large stones, known as megaliths, have been set in the ground in circular or square arrangements. Megaliths have been found near Rwa-sgrengs and Sa-skya in central Tibet, and in the far west at sPu, Shab-dge-sdings, gZhi-sde-mkhar, and Byi'u near Ma-pham lake. Close to Pang-gong lake in the northwest are eighteen parallel rows of standing stones aligned in an east-west fashion with circles of stones arranged at the end of each row. In western gTsang at Sa-dga' is a large gray stone slab surrounded by pillars of white quartz. Near Dang-ra lake are also large standing stones encircled by slabs, as well as sites that appear to be ancient square tombs. Western scholars have suggested these may be tombs or burial sites or possibly sacred arenas of some kind.

Unusual stones in striking shapes and colors are located in Kong-po, sPo-bo, and rTsa-ri, as well as caves and rock formations that have struck the inhabitants as having mysterious significance. The ancient kings of Tibet later erected stone pillars at their tombs and other locations in central Tibet. Whether these unusual stones and pillars are related in any way to the megaliths in other regions of Tibet is not clear.

The Tibetan megalithic sites have not been dated, but similar huge, undressed slabs of rock have been discovered in Europe at sites dated to as early as 3500 B.C. Traces of this megalithic culture have been found throughout the Mediterranean and on the east African coast. Megaliths have also been discovered in India and Southeast Asia, dated to about 1000 B.C.

Megaliths in Europe and south Asia are often found together with graves, and specialists propose that they are associated with specific religious ideas about an afterlife; other sites appear to be designed for the purposes of astronomy. The megalithic cultures in Europe are believed by some investigators to be connected to ancient Egyptian civilization, though some sites predate even Egypt. Someday persistent scholars may understand the significance of the Tibetan megaliths and know whether a relationship exists between the various kinds of Tibetan stones and the megaliths of other ancient civilizations.

OTHER EARLY INFLUENCES

Early influences on Tibetan culture may have come from the south and east in ancient times. Chinese archeologists see similarities between pottery found at the prehistoric Tibetan sites in Khams and pottery from regions in China. Discoveries of cowrie shells in Khams seem to indicate ancient trade with cultures around the Bay of Bengal or the Indian Ocean in this early farming period. These tantalizing finds call out for thorough investigations such as those undertaken in Europe and in other parts of Asia.

Now that archeologists have solid evidence that the plateau was inhabited in such ancient times, researchers will surely make the necessary efforts to fill in the huge gaps in Tibetan prehistory. Even today, however, history atlases can no longer show the Tibetan plateau as a vast blank marked "uninhabited" throughout the prehistoric era of world history.

How the earliest inhabitants are related to the culture which came to be known as distinctly Tibetan is an exciting question for future research. Tibetan history obviously does not begin with the Dharma Kings in the seventh century, or even with the days of the first king in the era before the beginning of the Christian era, for the roots of Tibetan culture extend back many centuries. Today few people are aware of the history of Tibet, but the story of Tibet's past will eventually take its proper place in the history of the ancient world.

PREHISTORIC ANCESTORS

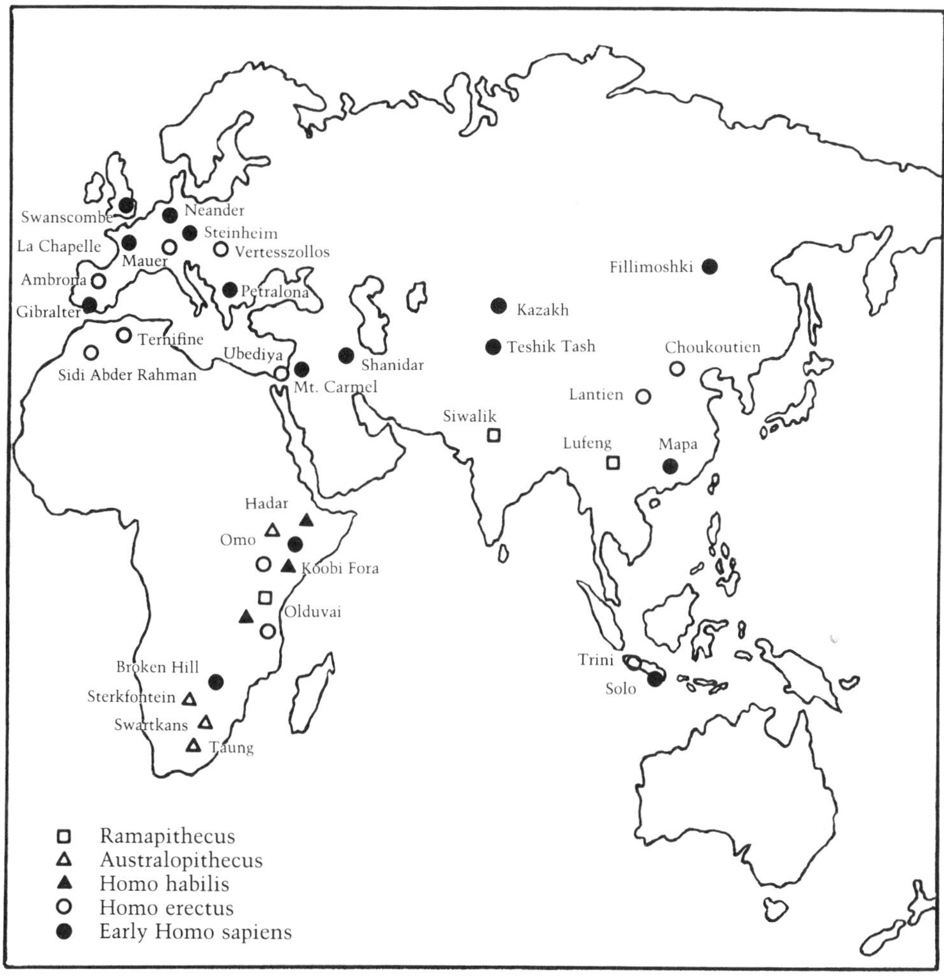

Sometime between 15 and 5 million years ago, the human line developed from ancestors related to the apes. *Ramapithecus,* who lived 15 million years ago, is no longer believed to be in the human line. *Australopithecus* lived 3.75 million years ago and appears to be a forerunner of *Homo habilis* and *Homo erectus,* true but primitive humans. *Homo erectus* flourished 500,000 years ago and presumably gave rise to several different types of early *Homo sapiens,* from which fully modern humans emerged about 40,000 years ago.

PREHISTORIC SITES IN TIBET

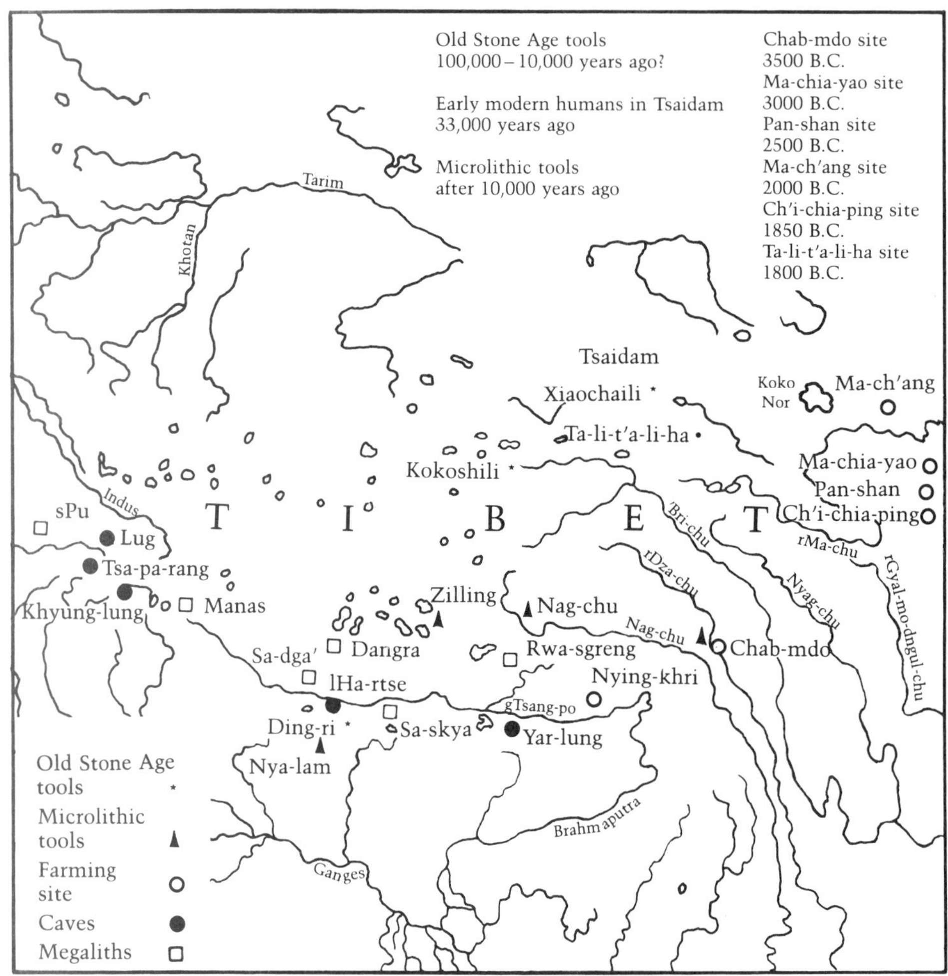

The most ancient prehistoric finds from Tibet are tools that are not yet precisely dated, but belong to the middle and late parts of the Old Stone Age. Tools from the middle of the Old Stone Age are generally associated with Neanderthal-type populations, while late Old Stone Age tools are related to early modern populations. Evidence of early modern man has been found in Tsaidam; hunting camps, evidenced by microlithic tools, and farming villages have also been discovered. Megalithic sites and caves are undated, but appear to belong to ancient times.

EARLY CULTURES AROUND THE WORLD

TIBET	EAST	WEST	AMERICA
12,000 B.C. Warming trend begins	11,000 B.C. Earliest known ceramic pots (from Japan)	8700 B.C. Goats, sheep herded in Iran	9000 B.C. Ancestors of American Indians arrive in N. America in large numbers
11,000–1000 B.C. Tibetan environment more favorable than at present: warmer, more humid, more trees	8000 B.C. Rice farming in Thailand	8600 B.C. Farming at Jericho: wheat, barley; first walled town	Beginning of overkill of large mammals
Microlithic tools in central Tibet may indicate hunting culture	7600 B.C. Fishermen from mainland settle in Japan	6800 B.C. First city: Çatal Hüyük in Anatolia; earliest woven cloth	9000 B.C. Hunter gatherer culture in Mexico, Central America
3500 B.C. Ancient dwellings in Khams	7000 B.C. Rice farming in S. China: spread from Indochina; water buffalo, and pigs	6700 B.C. Farming in S.E. Europe along Danube in loess soil; cattle kept	8000 B.C. Potatoes cultivated in Peru
3000 B.C. Farming villages in northeastern Tibet	5500 B.C. Millet farming in N. China along Huang Ho in loess soil	5000 B.C. Farming in Egypt and Mesopotamia	7500 B.C. Beans cultivated in Peru
2000 B.C. Farming villages in far northeast	6000 B.C. Farming in Indus valley, India	4500 B.C. Copper smelted in E. Europe; stone buildings in W. Europe	5700 B.C. Maize farming in Mexico
1850 B.C. Farming culture along Tao river	2900 B.C. Bronze Age Harappan culture in N.W. India	3500 B.C. Wheel invented in Mesopotamia	5000 B.C. Llama domesticated in Peru
Villages in Tsaidam, bronze in use	1800 B.C. Shang culture in N. China	3100 B.C. Egypt united by King Menes	3000 B.C. Old Copper Culture by Great Lakes N. America
1500 B.C. Ch'iang tribes west of Shang settlements near Ch'ang-an, N. China	1500 B.C. Writing in China	3000 B.C. Early Minoan culture in Crete	2100 B.C. Copper smelting in Andes
1000 B.C. Cooling trend: "New Ice Age"	1500 B.C. Bronze in use in N. China	3000 B.C. Sumerian urban culture; pictographic writing	1250 B.C. Olmec culture in Central America
Climate dries, forests decrease			

CHAPTER SIX

THE ORIGINS OF THE PEOPLE

The origin of the Tibetan people is one of the most complex research topics in ancient Tibetan history. The Bon tradition describes the Tibetan people as the descendants of gods or non-human nature spirits, and this accords with Bon religious tradition and cosmology. Other traditions trace the Tibetan people or their rulers to an Indian dynasty — a sophisticated and comprehensible explanation of Tibetan origins often favored by Buddhist historians. A tradition from the Maṇi bka'-'bum states that the Tibetan people descended from tribes of monkeys that developed gradually into human beings.

THE MONKEY ANCESTOR

According to the Maṇi bka'-'bum, a very intelligent monkey, blessed by Avalokiteśvara, grew more wise and compassionate. This unusual monkey attracted the attention of a brag-srin-mo, or rock demoness, who eagerly approached him, for her entire clan had disappeared. Though the monkey was not naturally drawn to her, friendliness and sweetness were his strongest traits, and he became her companion. Eventually, six monkey children were born; one being from each of the six realms of existence took birth as each of the monkey children, and thus the children had very different characteristics. The monkey's children rapidly multiplied, and soon there were some four hundred descendants. Caves behind Tse-thang on the mountain of Gong-po-ri are said to be the dwellings of these early inhabitants of Tibet.

As the population grew, the food the monkey's descendants collected in the forests gradually diminished until they were faced with starvation. In desperation, they began to fight with each other, and were finally separated by the monkey patriarch into four large tribes and two smaller groups. In this emergency, the old monkey was blessed once again with the creative inspiration of Avalokiteśvara. He realized for the first time that plants that could be used for food grew from seeds.

This gift of seeds he gave to the starving monkeys, planting the first fields of lentils, wheat, and several kinds of barley in Yar-lung. When the crops were ready, the monkey patriarch announced: "Now eat!" and so the place was known as Zo-dang. The monkeys enjoyed the new foods, and season after season they ate the grain. Slowly, understanding began to dawn upon them, and they too began planting some of the seeds instead of eating all of them. Eventually, there was more than enough extra food, and plenty of seeds to plant as well.

In addition to the gift of seeds, many blessings were bestowed upon Tibet to assure that in the future it would become the home of the Dharma and that its natural treasures would become resources to support the prosperity of the people. The tribes flourished, spreading into the valley and across the mountainsides.

FROM MONKEY TO HUMAN

The rGyal-rabs-gsal-ba'i-me-long explains that after generations of changes in their diet and their habits, the behavior and physical embodiment of the monkey's descendants were slowly transformed. The Chos-'byung Rin-po-che'i-gter-mdzod places the first changes much earlier, even before any seeds had been sown. They lost their fur and began to cover themselves with leaves and tree bark for protection. Language gradually developed, and distinctive customs emerged. Their interactions grew more complex, and concern for the welfare of the group as a whole increased.

The character traits of their parents were carried on among the tribes, so that some descendants were more like their father ancestor, others more like their mother ancestor, while some resembled both. The father's side was dominating and aggressive, but also reliable, intelligent, and compassionate, possessing an easily contented heart. Those resembling the mother were strong-willed, jealous, and very passionate. Sensuous and earthy, they enjoyed wealth, comfort, and food, and were more physically robust than the father's side. According to the Chos-'byung, when these early people recognized that beauty and grace were connected to virtuous behavior, they began deliberately to imitate beneficial practices. From these descendants of the monkey, the Tibetan people developed, spreading out from the Yar-lung valley across Tibet.

DATING THE MONKEY ANCESTOR

The account of the monkey's descendants is obviously similar to modern accounts of evolution that trace the human line back to ancestors related to monkeys and apes. But the modern theory of evolution is based on a timescale that encompasses many millions of years of prehistory, while the Maṇi bka'-'bum tradition is sometimes interpreted to refer to events in the time of Buddha Śākyamuni when the Tibetan land is said to have first become habitable.

According to the prediction in the Mañjuśrīmūlatantra, one hundred years after the Parinirvāṇa, the water covering Tibet will dry up, and the first forests appear. If this passage is taken to mean that the whole of the Tibetan plateau was covered with water, then it is hard to imagine where any monkeys might have lived until after the forests had developed. This would seem to fix a date for the emergence of the Tibetan tribes close to the time of the Buddha. But to what does this "drying of Tibet" refer?

THE "SEA OF TIBET"

We might imagine all of Tibet had been submerged beneath the sea, and in fact ancient shells have been found on mountain tops. But according to modern scientific understanding, the mountains were gradually pushed up from the ocean floor as continents collided. The sea shells and fossils reflect the early stages of mountain formation hundreds of millions of years before any humans walked the earth (see chapters 1 and 2).

But there are more recent periods during the last Ice Age, which began some three million years ago, when dramatic changes in the environment also took place. The water said to cover Tibet may not have been the ocean (rgya-mtsho), but rather numerous lakes (mtsho) that formed as glaciers melted during warmer periods in the Ice Age. As the plateau rose and grew increasingly dry, these lakes began to dry up (see chapter 3). Conceivably, the prediction could be understood this way. Or the drying may have occurred about 1000 B.C., a period of cooling and drying around the world. Thus parts of Tibet might have grown drier around the time of the Buddha Śākyamuni.

UNDERSTANDING THE MONKEY ACCOUNT

Connecting the development of the Tibetan tribes to the time of the Buddha, however, is more difficult. Tibetan historians offer various dates for the Buddha's Parinirvāṇa such as Atīśa's 2136 B.C., Sa-skya Paṇḍita's 2133 B.C., Phugs-pa-lhun-grub's 881 B.C., or Kha-che Paṇ-chen's 544 B.C. Whatever date we consider, we are dealing with a recent and short period of time into which we must fit all the events in Tibetan prehistory. Tibetan authors agree that after the development of the tribes and before the arrival of the first king, the land was divided among a series of twenty-five, twelve, and forty local rulers (see chapter 7). Though the dates of the first king also require careful consideration, it seems that he ruled sometime between the era of the Buddha and the fourth or third century B.C. (see chapter 10).

Though this might be a long enough period for the cultural developments described in the texts, there is certainly no time for the development of monkeys into human beings. Scientists today estimate that prehumans first emerged from the ape line about 5 million years ago, and modern man about 40,000 years ago (see chapter 5). To interpret the Maṇi bka'-'bum account as a foreshadowing of recent evolutionary theory may make it more convincing to modern readers, but then the monkey event needs to be placed in the distant past.

We have the option of making a special case for Tibet. Perhaps with the blessings of the Bodhisattva Avalokiteśvara, a group of unusual Tibetan monkeys rapidly developed in ways that elsewhere took far longer. This idea might be acceptable to many people in Tibet, both scholars and common folk, who regard the monkey as a manifestation of the great Bodhisattva. True, modern theories of evolution propose a human ancestor who was not actually like today's monkey, for the human line is thought to have descended from creatures that disappeared millions of years ago. But it would be difficult to say that Avalokiteśvara could not have taken whatever form was suitable.

If Avalokiteśvara was given the mission of guiding Tibet by the Buddha Śākyamuni, then there remains the question of timing. It may not be necessary to place this event in the recent past, for Buddhist scriptures make it clear that the compassionate actions of Buddhas have manifested in the world in many ages before Siddhartha Gautama became an Enlightened One. We may be uncertain how best to understand the Maṇi bka'-'bum tradition, and so prefer to label it a folktale. But an account from such a sophisticated source is worthy of careful consideration and cannot simply be dismissed.

OTHER ANCIENT INHABITANTS

Though the Maṇi bka'-'bum indicates that a new lineage of people began with the monkey's descendants, they may not have been the only or the earliest inhabitants of Tibet. If we look at archeological research, we find Old Stone Age tools over 35,000 years old (see chapter 5) made by some unknown population. Old Tibetan records such as the Lo-rgyus-chen-mo suggest that other inhabitants predated the first Tibetan tribes. These sources describe an early era of powerful and mysterious rulers. These rulers, known as the mi-ma-yin or nonhuman ones, apparently lived in Tibet for a long time before the Tibetan tribes finally took over the land. What became of them is never explained by historians, and perhaps some of these lines were continued in the Tibetan population.

The records about the mi-ma-yin are very fragmentary. We have no idea when they lived, how long any group was ascendant, or whether each group ruled the same territory. The first of these early rulers are known as gnod-sbyin-nag-po, "black demons." They used bows and arrows, but wore no ornaments. The name of the land at this time was bZang-yul-rgyan-med, Good Land Without Adornment. They were succeeded by "evil demons" known as bdud, who warred with one another using axes and battleaxes. They lived in the forested regions of river canyons and seem to have had a more complex culture involving rituals of some kind. The land was then known as the bDud-yul-kha-rag-rong-dgu, the Nine Valleys of Kha-rag Demon Land. After some indefinite period of time, power was in the hands of beings known as srin, more physically heavy and earthbound than other nonhumans. They used slings made of animal bones and catapult weapons, and the land was known as Srin-po-nag-po-dgu-yul, the Nine Lands of the Black Srin.

As the power of the srin declined, a fourth group of more godlike beings with a lighter, more positive quality gained control. Known as the lha, they are said to have used sharp swords. The land was known at that time as lHa-yul-gung-thang, God Land of the Central Plain. A group of very powerful beings with evil tendencies, known as the dmu-rgyal, were next in power. They used lassoes with hooks in warfare, and may have engaged in ceremonies and black magic rituals. They were followed by malignant creatures known as 'dre, who stayed in high places, rather than in the forested canyons. They used a bolo weapon made of ropes with rocks attached at the ends.

The seventh group are the Ma-sangs, a group of brothers or relatives, who are considered heroic and possibly humanlike in some ways. They were the first to use some form of armor and shields in warfare. They called the land Bod, the name used by the Tibetan people, for the name of the land is recorded as Bod-kha-nya-drug. These brothers included: gNyan-g·ya'-sbang-skyes, Gar-ting-nam-tsha, Gle-ngan-lam-tsang-skyes, Ru-tho-gar-skyes, She-do-ker-ting-nam, Me-pad-ma-skyes, gSang-ge-'phrul-po-che, Drang-ba-drangs-ma-mgur, and bKong-stong-nam-tsha.

After their rule, power shifted to the klu or nāgas, demi-god beings with great magical abilities. Capable of transforming themselves into many shapes, they are usually associated with a serpent form. There were said to be nine colonies of klu ruling Tibet, and the land was so called: Bod-khams-gling-dgu, the Nine Lands of Bod-khams.

The nāgas were followed by a mysterious group known simply as rgyal-po or rulers. It is not certain who they were—perhaps the survivors from all the previous eras, the strong ones who remained. During this period the land was known as Dem-po-tse. The tenth and last group before the tribes took power is known as 'gong-po, or enchanters. They were clever but savage, always misusing whatever good fortune came their way. The land was called sTong-sde-bco-brgyad, which might refer to the eighteen divisions of the sTong clan or to eighteen large districts.

THE MONKEY DESCENDANTS AND THE NON-HUMANS

The possible connections between the monkey's descendants and the nonhumans are rarely discussed in traditional histories. In listing the early stages of Tibetan culture, a few historians place the children of the Bodhisattva monkey just before the Ma-sangs brothers, while others place them after the last of the nonhuman rulers. But whether the monkey lineage became interconnected with the nonhumans or how this might have happened is not explained. Though the mother of the monkey's children is always said to be a demoness, historians do not discuss her ancestry. The two accounts are usually presented separately and left unrelated to each other.

Perhaps historians were assuming that the era of nonhuman rulers would show no connection with the development of human beings. Yet the nonhumans are always referred to as the rulers, so there must have been a population to rule. It may be that the two groups were living together, even intermarrying like the monkey and the demoness, and that the monkey's descendants gradually grew dominant or absorbed the older population.

An interesting legend recorded by R. A. Stein is told by tribes living near Tibet along the Min river east of rGyal-mo-rong. When they first arrived in this region long ago, these tribes encountered a primitive population they called the Qa. Heavyset with sloping foreheads, thick dark hair, and short tails, these savage inhabitants knew nothing of farming and lived in caves. After subduing the Qa with the aid of the gods, the newcomers began to call themselves the civilized people, the rMi.

This might suggest a way to fit the monkey descendants together with the nonhumans and the archeological evidence. We might consider the nonhumans a barbaric, primitive human population, and the monkey descendants a more advanced group, who eventually supplanted the older population. We might even speculate that different groups of mi-ma-yin were associated with the various ancient Stone Age tools and sites of habitation found on the Tibetan plateau.

But this would mean assuming that the Tibetan origin stories are really accounts of physical and cultural evolution from primitive to more civilized stages, as described by anthropologists. The situation, however, is more complex, for many nonhumans, such as klu and lha, are distinctly godlike rather than primitive. Traditional lists give eight different types of nonhumans, the sde-brgyad-mi-ma-yin. Buddhist texts describe numerous realms of higher space inhabited by godlike beings, each with specific characteristics. Beneath the god realms are lower, more physical realms where numerous other nonhumans reside. Even though these nonhumans are especially related to physical locations on the earth, their bodies are magical and their powers beyond those of ordinary humans. Bon texts, too, are familiar with a wide variety of nonhumans, and folk traditions recognize fine distinctions among different kinds of nature spirits. But accurate explanations of the character of different types of nonhumans are hard to come by, for the oldest material may have been distorted by later interpretations, and much information seems to have been lost. Now it is very difficult for us to know what the Era of Nonhuman Rulers really meant to the ancient Tibetans.

OTHER ORIGIN STORIES

A number of Tibetan sources have preserved traditions that trace the origins of human beings back to nonhuman ancestors rather than monkeys. Some Bon texts say the tribes descended from Srid-pa'i-lha Lum-lum and Chu-rgyal-mo-'bum. The Rlangs Po-ti-bse-ru speaks of six father gods, Yab-lha-drug, who arose from cosmic eggs. Among these father gods, the youngest was 'O-de-gung-rgyal, who married four different wives: a lha-mo or goddess, from whom the Tibetan kings arose; a gnyan-mo or demoness, from whose line all the variety of gnyan demons developed; a rmu-mo, another demoness from whom all the rmu demons descended; and a klu, or nāga wife from whom, many generations later, the human ancestor, A-mi Mu-zi-khri-do, arose. He in turn took three different nonhuman wives, a gnyan, a rmu, and a srin-mo, whose offspring gave rise to the Tibetan tribes.

An account in the La-dwags-rgyal-rabs tells of the origin of all beings from lha-srid, gods of existence. From one such lha, gShed-la-rogs, red-faced demons known as gnod-sbyin descended. From that line came two demons, one who fathered all the people of the regions surrounding Tibet, and the other who was known as brag-srin, rock demon. The rock demon gave rise to an ape demon, and his descendants spread through the borderlands. The brother of the ape demon gave rise to the ancestors of the Tibetan tribes.

We might speculate that the original tribes included certain unusual and powerful individuals referred to as the nonhumans. The most outstanding leaders from among the four original tribes were often referred to as the lha-rigs, the god lineage, and were renowned for psychic powers and other unusual traits (see chapter 8). Interpreting the nonhumans as unusual humans, however, does not explain the origin of the tribes. The lha-rigs or god lineage among the Tibetan tribes was sometimes said to intermarry with nonhumans, though the significance of such intermarriages is not explained clearly. But it seems that we cannot simply equate one group with the other.

MIXING OF LINEAGES

One theme seems to emerge from these various accounts. Even in genealogies that derive human beings from nonhumans, there are always several different kinds of nonhumans, some more physical, others more godlike. If we think in terms of the monkey and his demoness companion, there is clearly a mixing together of different lineages. And though both the monkey and the demoness had physical forms, their parentage was magical. Buddhist accounts explain that the Bodhisattva monkey took on a specific mission, for Avalokiteśvara had promised to guide the people of Tibet. In establishing a lineage of tribes that would continue into the future, the Bodhisattva monkey brought the blessings of Avalokiteśvara into the world of time and space.

Our time in history presents us with a certain outlook and shows us a particular version of the past. As a result, ancient ways of thought seem unconvincing to us. Yet modern forms of knowledge have a great deal of difficulty offering satisfying accounts of the development of a specifically human consciousness. It may be that in the beginning of human history, important events took place that we now lack the ability to understand. Perhaps if we could understand the older traditions that speak of such things, we might find deeper knowledge of human nature.

EVOLUTIONARY TIMETABLE

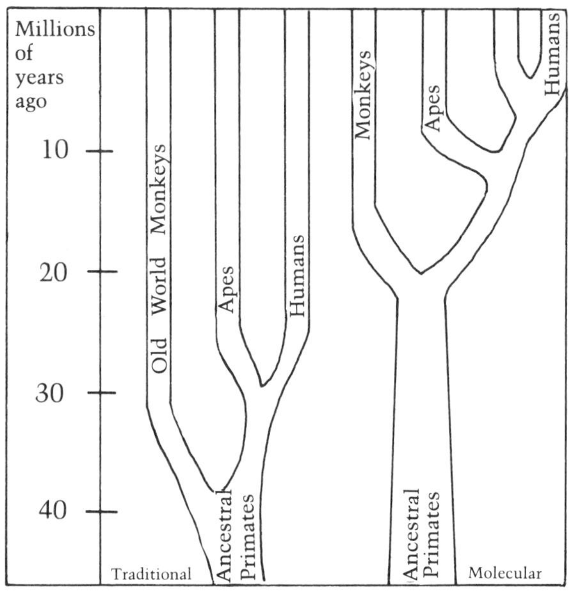

Different timetables for the emergence of the human line from the ape and monkey family are suggested by traditional anthropological methods and by molecular dating. This modern technique offers a measure of the closeness of relationship between organisms by determining the degree of similarity in proteins or DNA. This measure suggests that human ancestors emerged more recently than had been believed, but still more than five million years ago.

THE FIRST MODERN PEOPLE

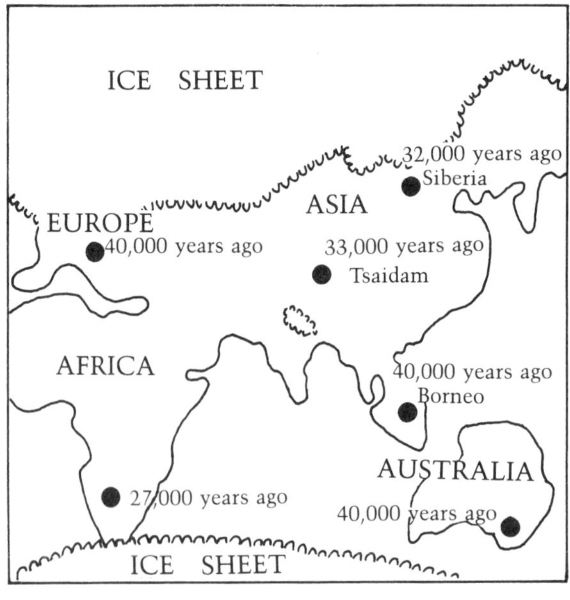

From ancestors that have been traced back 3.75 million years, modern humans first emerged 40,000 years ago, spreading rapidly and widely across Europe and Asia. Evidence based on genetic mapping suggests that modern humans might have first arisen west of Tibet and north of India. Early modern humans lived in the Tsaidam basin on the north edge of the plateau as early as 33,000 B.C. The connection between this population and later inhabitants is not yet clear.

THE EARLIEST INHABITANTS

TIBETAN CHRONICLES	SCIENTIFIC DATA	SCIENTIFIC DATA	SCIENTIFIC DATA
NONHUMAN RULERS: 10 groups who reigned in Tibet before the arrival of the 1st king: gNod-sbyin nag-po "Black Demons" bDud "Demons" Srin-mo "Ogres" lHa "Gods" dMu-rgyal "dMu lords" 'Dre "Evil demons" Ma-sangs "Nine brothers" Klu "Nāgas" rGyal-po "Rulers" 'Gong-po "Enchanters" RGYAL-PHRAN: Little kings who ruled over small territories; country is fragmented and threatened by neighbors 4th – 2nd c. B.C. gNya'-khri-btsan-po, first king of Tibetan dynasty, unites tribes	500 mya Earliest fossils from Tibet: shelled sea creatures 200 mya Tibetan plateau forming and rising out of sea as land masses join together; dinosaurs in forests of Tibet 45 mya India and S. Tibet join Asia, Himalayas form 25 mya Plateau slowly rising; forests cover much of Tibet, even Tsaidam and Byang-thang 10 – 5 mya Land drying out, lakes smaller, forests decreasing, but still plentiful; temperatures drop 3.75 mya Human ancestors evolving; *Australopithecus* in Africa 3.25 mya Beginning of Ice Age In glacial periods: Tibet drier, colder, less forested In interglacials: warmer, moister, more forested	2 million – 200,000 years ago Early Old Stone Age cultures in some parts of world *Homo erectus*, primitive man, in Asia, Africa, and Europe 500,000 years ago Tibet grows colder and drier; huge glaciers form 100,000 – 40,000 Middle Old Stone Age cultures in some parts of world Neanderthal man in Asia, Africa, and Europe Middle Old Stone Age tools from Tibet at Ding-ri (undated); Old Stone Age tools from Tibet at Kokoshili; uncertain era (undated) 40,000 – 10,000 Late Old Stone Age cultures in some parts of world Emergence of modern humans 33,000 Evidence of early modern humans in Tsaidam	12,000 B.C. Warming trend begins worldwide 10,000 – 5500 B.C. Tibetan climate grows more favorable; wider variety of plants and animals 10,000 – 5000 B.C. Beginning of finer, smaller tool technology of Middle Stone Age in some parts of world Tools of this style (undated) found in Tibet; megaliths, ancient cave dwellings (undated) 8000 – 4000 B.C. Beginning of farming in some parts of the world 5500 – 1000 B.C. Tibet warmer and more forested than at present 3500 B.C. Ancient dwellings in Khams 3000 B.C. Farming in far northeast Tibet 1000 B.C. "New Ice Age:" Tibet's forests decrease; climate colder, drier

mya = millions of years ago

CHAPTER SEVEN

THE ANCIENT TRIBES OF TIBET

The ancient Tibetan people were divided into a number of distinct tribes. Four great tribes are mentioned by the bKa'-chems-ka-khol-ma, while the rGyal-rabs-gsal-ba'i-me-long and the Maṇi bka'-'bum speak of six tribes. Names of the original tribes can be found in many histories: bSe, lDong, dMu, and sTong, together with dBas and Zla or sBra and 'Bru.

The ancestors of the tribes are said to be the six children of the monkey (see chapter 6). There is, however, no generally accepted way of connecting the monkey descendants with specific tribes. It is not clear that each of the six children became an ancestor of a tribe. Some histories describe how the growing population of monkey descendants was only later separated into groups that became the tribes, while other texts do not integrate the account of the monkey with the tribes at all.

Old records offer a number of explanations about the division into four great tribes, the rus-chen-bzhi, and into the six tribes, the mi'u-gdung-drug. Rus refers to the father's side of the family, called the "bone" line, while the mother's side is known as the "blood" line. Sometimes instead of four tribes, five are spoken of, rus-chen-lnga, with a zhangs-po-lha-rigs rGo, the uncle-god lineage of rGo, being added to the four. This lha-rigs was especially connected to the gods and spirits, for its members were very devoted to ritual and to maintaining ceremonial relationships to natural powers. Virtuous, obedient, and respectful, this tribe was also known as the tsha-zhang, nephew-uncle line, and was renowned for psychic powers. Sometimes the lha-rigs is said to be made up of the "good" leaders of each tribe. These outstanding clans are: bSe-khyung sBra; dMu-tsha rKa; A-spo lDong; and sTong A-lcags 'Bru.

Another way of dividing the tribes distinguishes the four "good" tribes from the less distinguished, who form two groups, the dBa' and the Zla, to give a total of six. If the zhangs-po-lha-rigs rGo is added, then seven are counted. On the other hand, the fifth group, the dBa', are considered by some sources to be the lha-rigs, while others say the dBa' is the tsha-zhang. dBa' signifies powerful and mighty, and by contrast, the Zla are sometimes regarded as a secondary or inferior group. Zla was also known as mang, common, and regarded as the "bad younger brother" of the other tribes.

The tribes can also be related to specific parts of the body by using Chinese calculation techniques, known as the Golden Tortoise system, brought into Tibet during the 7th–8th centuries by two Chinese princesses. Each tribe is also associated with different qualities and with various elements as well as with animals, which are symbols of the tribe's bla or spirit. The dMu tribes are associated with wood and with sheep; the rGo lineage with fire and the goat; the lDong lineage with earth and antelope; the bSe with iron and the kiang or Tibetan wild ass; and the sTong with water and the yak.

LITTLE KINGDOMS

As the power of the nonhumans declined, the original tribes took control of the land, though systematic histories never describe any specific events in this process. In various chronicles scattered remarks can be found about struggles among the Ma-sangs, nonhumans, and certain tribes in ancient times. It would be interesting to assemble and compare such references.

Eventually twelve, twenty-five, and then forty small kingdoms arose, about which we have very little information except for some of the names of rulers and regions. Lists from Tun-huang and other old records include such places as mChims-yul, Zhang-zhung, Myang-ro, gNubs-yul, Myang-ro-sham-bong, sKyi-ro/Gyi-ri, Ngas-po/Ngam-shod, 'O-yul/'Ol-phu, Sribs-yul/Srin-yul, Kong-yul, Myang-yul/Nyang-yul, Dwags-yul, dBye-ro/gYe-na, Klum-ro, Sum-yul, 'Brog-mo, and rNgegs/rNgegs-yul Se-mo.

Tibetan histories describe how during the centuries before the first king, power was divided among these many factions. One group or another was dominant for a while, only to decline or be overthrown. In this era of fragmentation, according to the bKa'-thang-sde-lnga, time and again Tibet was threatened on all sides by powerful neighbors, while within its borders there was no regard for moral behavior. Though laws existed to punish wrongdoing, people lacked concern for society as a whole, and were driven by personal greed for territory and property. No ruler possessed enough strength or respect to unite the little kingdoms and protect the land. At the mercy of confusion and poverty, the population could not progress beyond the level of bare survival.

DATING THE LITTLE KINGDOMS

The timeframe for these small kingdoms is an open question. No dates are given in histories, though it is generally said they were in existence before the time of the first king, who united them in the third or fourth century B.C. (see chapter 9). Several accounts from Tun-huang agree that these small kingdoms flourished before the time of the first king. The lists of small kingdoms seem to have been deeply embedded in the culture, for one Tun-huang document shows they were used in ceremonies during the enthronement of a new king to reaffirm the right to reign over traditional territories ruled by the first king.

But there is evidence suggesting that some of these kingdoms were not controlled by the descendants of the first king. Tun-huang records indicate that Zhang-zhung and Ngas-po came under royal control only in the time of Srong-btsan-sgam-po and his father in the sixth and seventh centuries. The relation between specific little kingdoms and the Tibetan rulers in earlier times is hard to trace for a number of reasons. There may have been a loose confederation of little kingdoms that shifted allegiance over the centuries as the political situation changed. Tun-huang records show that kingdoms were sometimes renamed as power changed hands, and that some regions were conquered and reconquered several different times.

Lists of the little kingdoms name the rulers of each principality, and these names show up in other accounts as well. But some are found in stories about ancient times while others are mentioned in the records about the later Dharma Kings. Some scholars take this as evidence that at least some of the little kingdoms were no older than the sixth or perhaps fifth centuries. Other scholars have proposed that these names might be titles handed down from generation to generation, and this seems a reasonable possibility.

This subject would be worth investigating more deeply, for each kingdom seems to have had distinctive traditions and may have made valuable contributions to the development of Tibetan culture. A number of documents from Tun-huang and other Central Asian sites might shed more light on these principalities. Some of these documents, for example, offer lists of kingdoms, with their capitals, rulers, queens and ministers, together with important local spirits and shamans.

The peoples, or at least the rulers, of these little kingdoms belonged to the Tibetan tribes, for tribal names—for example, sTong, Khyung, dBas, and bSe—can be recognized in the lists of rulers and ministers. A great variety of clan names are also given; some, such as Rlang and mGar/'Gar, are well-known, but others, such as rNgo, Kam, and Su-ru, are less familiar.

Other lists of important early tribes that differ from the traditional six can be found in accounts describing the people who met the first king. It is commonly said that the new king, who united the twelve kingdoms, was first greeted by twelve people. But it is not at all certain that the twelve individuals were from the twelve kingdoms. Some texts mention two people from each of the six tribes: the bTsan-pa lHo and gNyags, the bTsun-pa Khyung and sNubs, the gNyan-pa Se and sPo, together with lHa-bo-lha-sras, Se-bon, rMa-bon, Cog-la-bon, Zhang-zhung-bon, and Tshe-mi-bon—perhaps magicians or shamans. Other texts say twelve people who all came from Sum-pa greeted him.

Among the Tun-huang documents is another list of important early tribes: lHo and rNgegs, Sha and sPug, mTshe and gTso. This group is known as the 'bangs-rus-drug, the six clans of royal subjects. Bon texts sometimes give still other lists of ancient tribes: Phya, dMu, Tsug, and Go. The relation among these different versions and better-known lists of the four or six great tribes needs to be determined.

FOUR GREAT CULTURES

At some point in time, the tribes were connected with four great cultures: bSe with 'A-zha, dMu with Zhang-zhung, sTong with Sum-pa, and lDong with Mi-nyag. These cultures may have been founded by people with ancestry different from the Tibetans, becoming associated with the tribes at a later date. Or they may have been established by the Tibetan tribes, who over time developed distinctive cultures in different regions.

According to some Bon texts, the four tribes, the twelve little kingdoms, and the four cultures were linked together even before the first king. According to these sources, the twelve kingdoms sent tribal representatives to the king: lDong from Mi-nyag, 'Bru from Sum-pa, sBra from Zhang-zhung, and rKa from 'A-zha. But other records paint a different picture. For example, in some Tun-huang documents the bSe tribe is called one of the ancient vassals of the king in Srong-btsan-sgam-po's father's time, while other Tun-huang records clearly point out that 'A-zha was first conquered by Srong-btsan-sgam-po. The names of Zhang-zhung and Sum-pa can be found in the lists of the twelve little kingdoms, but the names of 'A-zha and Mi-nyag are not found.

From the Tun-huang documents we know approximate locations for some of these cultures in the time of the Dharma Kings, for these cultures were part of the Tibetan Empire at that time. 'A-zha was located in the northeast around Koko Nor, Zhang-zhung was centered around Manasarowar in the far west, Sum-pa occupied parts of the east and the central plains of Byang-thang, and Mi-nyag was along the eastern borderlands. How extensive these cultures might have been in more ancient times is uncertain.

To clarify these issues, one would need to sift through many different kinds of material—genealogies, biographies, local records, Bon lore, and folk stories—each of which presents special difficulties. Origin stories exist about some famous clans, but often these are juxtaposed to seventh or eighth century events, leaving huge unexplained gaps in chronology. Lists of ancestors and descendants are hard to date without cross-referencing individuals, tracking them through a maze of aliases and nicknames. Folk stories sometimes mix together historical events with strange episodes that are more difficult to evaluate.

Nonetheless, a thorough study of the history of the tribes might help resolve some interesting questions. Did the tribes originally have specific territories? Did they speak different languages or dialects? The Zhang-zhung culture is said to have had distinctive language in ancient times, and some regions of modern Tibet, such as Hor-khog, rGyal-mo-rong, and Mi-nyag, are known to have languages that differ from Tibetan. Did each tribe have its own distinctive sacred rituals and oral traditions (as certain rituals are said to be especially connected with the Sum-pa culture even today)? Could different oral traditions from the various tribes help explain the discrepancies sometimes found in comparing old records and genealogies?

NEIGHBORING PEOPLES

When the connections between the original great tribes, the little kingdoms, the four cultures, and the important ancient clans are more fully understood, it might be possible to determine if the original tribes included most of the people later known as the Tibetans. Tibetan historians agree that the tribes are the ancestors of the Tibetan people, but the claim that the tribes are the only ancestors may be too broad. Even when we can confidently follow a lineage back from son to father for many generations, we are still disregarding the mother's side of the family, which might reflect intermarriages with neighboring peoples. A careful study of numerous genealogies could reveal when such intermarriages had taken place.

How the Tibetan ancestors might be related to the inhabitants of neighboring regions needs to be carefully considered. Some genealogies link neighboring peoples together with the Tibetans through common ancestors that emerged at some undetermined point in the past. These lineages then continued their separate ways, giving rise to people of Tibet, Kashmir, Sa-hor, Persia, and so forth. Unless we are willing to interpret the monkey tradition to mean that Tibet was the original homeland of all human beings, it seems clear that people with ancestries different from the Tibetans might have been living on the Tibetan plateau.

CH'IANG TRIBES

A number of old Chinese annals suggest connections between Tibetan tribes and peoples reported to be living west of China in the Han dynasty (202 B.C.–220 A.D.) histories. The ancient Chinese historians called these tribes the Ch'iang and described them as hunters with a reputation for bravery. Han dynasty records contain accounts of the Ch'iang's activities in more remote times, noting that their homeland had originally been far to the west, and that they were raiding the farming settlements near Ch'ang-an (Xian) as early as 1400 B.C. The Ch'iang are mentioned in the oldest specimens of Chinese writing inscribed on animal bones used in divination. Many such bones have been discovered at Anyang in northern China, and dated to about 1500 B.C.

Han dynasty annals note that by 500 B.C. the Ch'iang were organizing into tribes and had begun farming and raising cattle. They were reported all along the borderlands west of China by the second century B.C., and had developed into one hundred and fifty tribes in the second century A.D., while seventh century records note over fifty scattered tribes.

Some T'ang dynasty (618–907 A.D.) records say that the seventh century Tibetans, whom they called the T'u-fan, lived in a land that had been Western Ch'iang country, but the annalists remark that they do not know the T'u-fan's original tribal affiliation. Other T'ang accounts state that the Ch'iang were the ancestors of the T'u-fan, whom the T'ang Empire first encountered in the seventh century when the Tibetan King Srong-btsan-sgam-po sent his emissary to their court.

While modern research has generally found the old Chinese histories to be reliable sources for dates and certain kinds of information, the interpretations of these ancient historians may well have been influenced by their own cultural perspective. For example, how did they ascertain which of the tribes west of China were Ch'iang, and what criteria did they use—ethnic, linguistic, territorial, or cultural?

It is also obvious that information available to these annalists may not have been complete, especially about inhabitants of remote regions. T'ang historians themselves remark that they had not heard of the Yar-lung dynasty of Tibetan kings in central Tibet until the time of Srong-btsan-sgam-po in the seventh century. Many facts must have been gathered by hearsay, perhaps from the inhabitants of borderlands between China and Tibet.

Whether or not the T'ang annalists' opinion about the ancestors of the Tibetans is correct, the Ch'iang theory raises important questions about the composition of the early tribes. A few tribes of Ch'iang still remain in borderland areas even today, and some scholars see similarities between their language and culture and those of the Tibetans. Some of the Ch'iang tribes may have been among the little kingdoms, but until more research is done, it would be premature to draw conclusions.

"TIBETO-BURMESE" LANGUAGES

Modern linguists see similarities among Tibetan, Burmese, and numerous languages in the Himalayas and along the eastern edge of the Tibetan plateau. Some scholars have suggested that Tibetan and Burmese ancestors were part of a group of ancient peoples speaking "Proto-Tibeto-Burmese." This kind of speculation is not unlike theories about the origins of the modern speakers of Indo-European languages. But as yet linguistic theories have not been corroborated by the Tibetan historical tradition or records of other peoples.

One such theory discussed by S. K. Chatterjee suggests that long before 1000 B.C., peoples ancestral to the ancient Tibetans and speaking dialects of "Tibeto-Burmese" were centered somewhere to the east of Tibet and north of Burma. As various groups migrated west into Tibet and south into Assam and the northeastern corner of India along the Himalayas, distinctive but related languages emerged.

It is interesting that a number of modern historians of India such as Basham consider the possibility that these old Himalayan kingdoms were non-Aryan. Chatterjee suggests that "Tibeto-Burmese" tribes may have founded some of the dynasties located in the foothills, such as the Koliyas, the Vṛjjis, the Śākyas, and the Licchavis, gradually coming to speak Indian languages. Sylvain Lévi noted that the so-called kirāta peoples in the ancient Vedic tradition seem to have been non-Aryan. These people, described as fierce warriors with golden complexions, lived in the northeast hills and by the Bay of Bengal.

It seems reasonable that there could have been close links in very early times among the ancestors of the Tibetans, the Burmese, and the peoples in the Himalayas and the eastern borderlands. But linguists have pointed out that it is very difficult to know how to interpret the similarities found among languages in these regions, and some suggest that the theoretical framework for comparative study of "Tibeto-Burmese" languages is still inadequate. Studies of Tibetan and southeast Asian languages are not as old and well established as research in Indo-European languages. But one day linguists may be able to show us more about the relationship between the Tibetan people and their neighbors.

THE NAME BOD

At some point in the history of the tribes, the kingdom began to be known as Bod. Some modern scholars believe that the word "Bod" has a specific meaning and relate it to various Tibetan verbs such as "bon" or "'bod." The Blue Annals notes that old records show an earlier name was sPu-rgyal, but that even in the time of the Buddha, the land was called Bod.

The name Bod shows up in old records of the names of the land before the days of the little kingdoms. Trying to trace the usage of the name and its relation to sPu-rgyal might reveal more information about the early tribes.

Neighboring countries have had different names for Tibet at different times. In T'ang China it was called "T'u-fan" and "Tu-po." How these names were actually pronounced is not certain. The ninth and tenth century Arabs called Tibet "Tubbat," while the Indians have long referred to it as "Bhoṭa." "Tibet" seems to have come from the "Tubbat" of Moslem authors.

Given more time and effort, researchers may eventually be able to synthesize historical, linguistic, and even archeological data to create a vivid and accurate portrait of the ancient Tibetans and their relationships to other peoples in Asia. If Tibetan historians could contribute a more complete understanding of Tibetan traditions about the ancient past, research might move beyond interpretation to discover more solid facts about the early development of the Tibetan people.

LITTLE KINGDOMS OF ANCIENT TIBET

Before the first king of Tibet, gNya'-khri-btsan-po, arrived in the 3rd century B.C., Tibetan tribes were not united. Constant struggles took place among the tribes, who were ruled by chiefs holding power only within their own regions. Traditional lists of these kingdoms include twelve or more place-names, not all of which can be located with certainty. A number of kingdoms seem to have been centered around the gTsang-po and the Nag-chu rivers in central Tibet.

CH'IANG TRIBES ON THE PLATEAU

According to Han dynasty annals, tribes known as Ch'iang and Ti or Tik lived west of Ch'ang-an as early as 1500 B.C., and were reported all along the eastern borderlands of the Tibetan plateau by 500 B.C. They began spreading north and south as they divided into many tribes. Some T'ang records call the Ch'iang the ancestors of the Tibetans, but not until the seventh century did Chinese historians have any knowledge of central Tibet where the ancient kings resided.

RGYAL-PHRAN:

LITTLE KINGDOMS OF ANCIENT TIBET

KINGDOMS, RULERS, AND MINISTERS

mChims-yul Gru-shul
mChims-rje Gu-yod
Blon-po sNang and Ding

Zhang-zhung-yul
rGyal-po Lig-snya-shur
Blon-po rMa and Dar-sang

Myang-ro-phyong-dkar
gTsang-rje Thod-dkar
Blon-po gNang

gNubs-yul-gling-dgu
gNubs-rje dMigs-pa
Blon-po rMe'u and 'Bro

Nyang-ro-sham-bong
rNgam-rje 'Brom
Blon-po Shes and Sug

Gyi-ri-ljongs-sngon
Gyi-rje rMang-po
Blon-po Ngam and 'Bro

Ngam-shod-khra-snar
Zing-rje Khri-'phrang-gsum
Blon-po mGar and gNyan

'Ol-phu-spang-mkhar
Zing-rje Thon-greng
Blon-po rNgog and sBas

Srin-rong La-mo-gong
Brang-rje Gong-nam
Blon-po sBrad and Zhugs

Kong-yul-bre-sna
Kong-rje Dar-po
Blon-po Tug and mKhar-ba

Nyang-yul-rnam-gsum
Nyang-btsun Glang-rgyal
Blon-po Bug and Do-rje

Dwags-yul-gru-bzhi
Dwags-rje Mang-po-rgyal
Blon-po Rlangs and sKam

'Brog-mo-rnams-gsum
rGyal-po Se-mi-ra-khrid
Blon-po Ka-ra-nag

gYe-mo-yul-drug
gYe-rje mKhar-ba
Blon-po rBo and sTug

Se-mo-gru-bzhi
gNyags-rje Gru-'brang
Blon-po Se and Nyang

CHAPTER EIGHT

DESCENDANTS OF THE TIBETAN TRIBES

The history of the Tibetan tribes is the actual history of the Tibetan people. Unfortunately, Tibetan authors often reduce the history of the tribes to a few lines, and quickly pass on to the arrival of the first king from a foreign land centuries later. American histories also used to begin with the exploration and founding of colonies by foreigners. But unlike the American Indians, the Tibetan tribes were not supplanted by a new population whose own history and culture were naturally of greater interest to their own historians. Somehow, in the eyes of most Tibetan authors, the entry of a single king into the fabric of Tibetan culture made centuries of history seem unimportant.

Since the early history of the tribes is not yet well researched, we have assembled only an outline. The Po-ti-bse-ru, the 'Gu-log Chos-'byung, the lHo-rong Chos-'byung, and various gto-chogs may eventually clarify the tribes' histories.

THE BSE TRIBE

The bSe tribe is led by the clan known as bSe-khyung sBra, and is famous for great wealth. The Po-ti-bse-ru traces bSe back to the gNyan wife of A-mi Mu-zi-khri-do (see chapter 6), while other sources give the first bSe ancestor as Gling-ser-thang-rje, a descendant of the srid-pa'i-lha. The sBra are generally divided into three groups: nine black, nine white, and nine mixed. Some sources connect the white sBra with Zhang-zhung, sPu-rang, and Bal-po in the west, the black sBra with the eighteen tribes of rGyal-mo-rong in the east and with 'A-zha in the far northeast. rGyal-mo-rong districts include lCag-la, Wa-si, Khro-skyab, So-mang, rDzong-'ga, lCog-rtse, Dam-pa, 'A-gzhi, mDo-li, Bra-sti, Ba-bam, dGe-shi-rtsa, Hwa-hwa, Len-tsa, Rab-brtan, bTsan-la, rGyal-kha, and mGron-bu. The mixed sBra includes the Ar or A-dra lineage and is linked to central Tibet. This line connects to the Rlang clan, which in turn gave rise to the mGar clan, according to some sources (though others derive mGar/'Gar from the lha-rigs rGo).

In the ancient times of the twelve little kingdoms, we can find a member of the Khyung-po clan of the bSe tribe listed as the minister to the region of Zhang-zhung, while the Rlang clan was connected to Sum-yul. Some lists of clansmen meeting the first king include both Khyung and bSe.

Tun-huang documents show that by the sixth century, in the days of King gNam-ri, Khyung-po became a very powerful clan that provided the king with a famous minister—Zu-tse, a great military leader, who conquered part of rTsang. In the days of the Dharma Kings, Khyung-po held lands in rTsang, and both Khyung-po and 'Bro, a clan that provided wives for several kings, were responsible for the military troops in Ru-lag, a region in gTsang (which seems to be rTsang). Khyung-po was the clan to which two famous later Buddhist masters belonged, Mi-la-ras-pa and Khyung-po rNal-'byor.

The mGar clan also provided ministers for the kings beginning in the time of Srong-btsan-sgam-po's father. The famous mGar sTong-rtsan-yul-gzung served King Srong-btsan-sgam-po. A descendant of mGar, A-mnyes Byams-pa'i-dpal, was a disciple of Padmasambhava. He traveled to Khams to settle in the region of Gling, and from his line came the founders of sDe-dge.

A-mnyes Byang-chub-'dre-'khol, a great siddha in the Rlangs lineage and famous as Ge-sar's teacher, was also in Gling shortly after this time. Among his descendants was the famous T'ai Si-tu Phag-mo-gru. Other famous descendants of the Rlangs ancestors dPa'-bo and Rlangs lHa-gzigs were Rlangs Khams-pa Go-chas, who was one of the first Tibetan monks, Rlangs dPal-gyi-rdo-rje, and Rlangs dPal-gyi-seng-ge, a disciple of the Abbot Śāntarakṣita. In Khri-srong-lde-btsan's time, a daughter of Rlangs Lo-tsā-ba married dPal-po-che, a member of the 'Khon lineage. Their descendants included the great 'Khon dKon-mchog-rgyal-po (1034–1102) of Sa-skya. Rlangs is also said to have connections with Kashmir and China.

The connection of the bSe tribe with rGyal-mo-rong and 'A-zha in the east is harder to trace, for the histories of these two regions are not well known. It is difficult even to say how old the culture of 'A-zha might be. It is not listed among the ancient twelve little kingdoms, but Tibetan historians sometimes mention 'A-zha in connection with Bon-pos who arrived in Tibet during the reign of the eighth king Gri-gum. Thus it seems to belong to early times. 'A-zha is also named as the home of doctors who helped heal the eyes of King sTag-ri-gnyan-gzigs, the grandfather of Srong-btsan-sgam-po. We might speculate that these so-called black clans of sBra were specialists in magic particularly connected to the Bon tradition.

The Tun-huang documents about 'A-zha mention several unusual clan names, such as Mag-ga/Meg-le, Da-red, and Mug-lden, and others more familiar, such as 'Bro, Cog-ro, Shud-pu. But there were many Tibetan officials in 'A-zha in the days of the Dharma Kings, and it would be hard to tell which names might be distinctively 'A-zha. It seems that 'A-zha was not well integrated into the Tibetan kingdom until the time of Srong-btsan-sgam-po. The Tun-huang annals note that this king was the first to subjugate 'A-zha. But its connection with the bSe tribe might be older than this. Srong-btsan-sgam-po's son Gung-srong married an 'A-zha princess, who was the mother of Prince Mang-srong. It is interesting that the minister mGar spent a number of years helping reorganize 'A-zha as part of the Tibetan Empire after it was completely conquered in 670. Perhaps the king had chosen him because he had connections with the 'A-zha clans. Later rulers of Mi-nyag also came from bSe (see below).

THE LDONG TRIBE

The lDong tribe, connected with Mi-nyag, is led by the clan known as A-spo lDong, and is famous for its great leaders. The Po-ti-bse-ru also traces lDong back to the gNyan wife of A-mi Mu-zi-khri-do, while other texts name the lDong ancestor as sKu-rje-khrug-pa, a descendant of the srid-pa'i-lha. There are eighteen branches including common people, royal tribes, minister tribes, and nobles. Though ministers from lDong are not recorded for the earliest kings, some lists of clansmen meeting the first king include a clan called sPo, which might be connected to the leading clan of lDong, the A-spo lDong. Bon sources note that the lDong flourished because they performed important rituals in the days before the first king. Ancient connections exist between the lDong and Sa-skya, and some sources describe an old rivalry between the ancestor of the 'Khon clan and the eighteen tribes of lDong in the time of the Ma-sangs brothers, long before the era of the first king.

By the sixth century the tribes of lDong and sTong seem to be associated closely with each other, according to the Tun-huang records. But their locations are not certain. The lDong are also connected with Rlangs in the days of the Dharma Kings, twelve generations before A-mnyes Byang-chub-'dre-'khol's father, when the Rlangs ancestor dPa'-bo married a wife from lDong.

The dBas clan is sometimes connected to the lDong and sometimes considered a separate tribe. Some texts connect it to the rGod lDong, noting that members of the dBas were among the troops stationed in the far north in the era of the Dharma Kings. dBas are often characterized as physically powerful warriors, athletic heroes with almost magical abilities. A member of the dBas clan was put in charge of the cavalry of the Tibetan army by Srong-btsan-sgam-po, and dBas held lands called Zha-gang-sde-gsum in Khams. In this same era, members of this clan were positioned along the borders with 'Jang (Nan-chao) southeast of Tibet. They provided ministers and wives for the Dharma Kings, especially after the mGar family fell from power about the beginning of the eighth century.

The famous religious master rDo-rje-rgyal-po of Phag-mo-gru (1110–1170) belonged to the dBas We-na-'phan-thog clan through his father, while his mother was sister of the mother of Kaḥ-dam-pa bDe-gshegs (1122–1192), the founder of Kaḥ-thog monastery, whose family belonged to the sGa clan.

In the ninth and tenth centuries, the division of lDong known as lDong sPom-ra or the lHa lDong-dkar-po was connected to the lineage of Tibetan kings through King 'Od-srung's line, which included dPal-'khor, bKra-shis-brtsegs, 'Od-lde, and Khri-lde (see chapter 20). Other connections exist between the lha-rigs lDong and Sa-skya, with 'Bri-gung, Bla-ma Zhang's line, Mar-pa's family, and the Rlangs clan. Ge-sar is generally said to have belonged to the lDong-smug-po clan.

lDong is also often associated with the lha-rigs rGo/sGo tribe. According to old records, the rGo lineage spread widely throughout Khams, and included the 'Gar clan, as well as Ke, 'Gol, gSung, gSer, 'Brom, Chi, 'Bu, bZhag, Shol, sTag, 'Pyung, gZhe, Seng, Ram, and the three Phyugs clans.

Although the tribe of lDong is usually associated with Mi-nyag, the tribe seems older than Mi-nyag, which is not listed among the ancient twelve little kingdoms. But the history of Mi-nyag is not yet well researched. We know that Srong-btsan-sgam-po married a princess of Mi-nyag, but nothing much is known about her.

In later times, a kingdom of Mi-nyag existed in the northeast beyond Koko Nor, where the older 'A-zha kingdom had been located. This 11th–12th century Mi-nyag seems to have become part of the Hsi-hsia kingdom, for Tibetan historians note that Genghis Khan died in Mi-nyag, which matches foreign accounts of his death in Hsi-hsia. The rulers of the northern Mi-nyag traced their lineage back to one Se-hu, who ruled 260 years before the Hsi-hsia kingdom was conquered by Genghis Khan in 1226. Hsi-hsia became a powerful kingdom, warring with Sung China, the Uighurs, and the inhabitants of Tsong-kha in the eleventh century. The Hsi-hsia language is said to be related to Tibetan, and Tibetan was the religious language of the monasteries.

After the fall of Hsi-hsia, descendants of these kings settled in gTsang-gnam-ring. Several centuries later, the father of the famous siddha Thang-stong-rgyal-po was said to have come from the lHa lDong-dkar-po of Mi-nyag. Some sources note that the lineage of the first king of Sikkim in the seventeenth century, Phun-tshogs-rnam-rgyal, was connected to Mi-nyag.

But there is another Mi-nyag in the southeast around Dar-rtse-mdo and lCag-la, where the snow-capped peak Mi-nyag-gangs-dkar (Minya Konkar) is located. The history of this southern Mi-nyag is even less well known. Mi-nyag in the southeast and 'Jang, the land south of Mi-nyag around Ta-li lake, may be connected to the lDong through the rGo lineage. Some branches of rGo are said to have originated near Mi-nyag-gangs-dkar when a heavenly being descended upon the mountain.

THE STONG TRIBE

The sTong tribe is led by the clan of A-lcags 'Bru, and is famous for its great heroes. The Po-ti-bse-ru also traces sTong to the gNyan wife of A-mi Mu-zi-khri-do, while other sources give the first sTong ancestor as Ring-rje-'u-ra, a descendant of the srid-pa'i-lha. sTong has eighteen branches and is closely connected with Sum-pa, a very important culture in most ancient times. Sum-yul is among the ancient twelve little kingdoms, and O-yong-rgyal-ba, a shaman from Sum-pa, apparently mounted an unsuccessful challenge against the first king (see chapter 9). Some sources claim that the twelve men who met the first king all came from Sum-pa.

The sTong tribe provided ministers for the early kings known as the Eight lDe, and women from the sTong clan married several kings, including lHa-tho-tho-ri's father, and sTag-ri, the grandfather of Srong-btsan-sgam-po. Tun-huang records show that the sTong were considered ancient allies of the Yar-lung dynasty, while the Sum-pa peoples were known to be part of the Tibetan kingdom in the sixth century in the time of Srong-btsan-sgam-po's father, gNam-ri.

In the time of Srong-btsan-sgam-po, the Sum-pa lands seem to have been located around Nag-chu-kha and included 'Jong, 'Dre, and rGod-tshangs, while the 'Bru clan of the sTong tribe was especially connected to Klum-ro and Nam-po toward the northeast. In the eighth century, Sum-pa tribes led by a famous general Zhang sTong-rtsan were sent to the far northeast (to the region later known as Mi-nyag) to guard the frontier. This may help explain the close connection between sTong and lDong.

Branches of sTong also gave rise to the sKyu-ra clan to which the father of 'Bri-gung Rin-po-che 'Jig-rten-mgon-po (1143–1192) belonged. There are also connections with the 'Khon clan of Sa-skya, the rGyal-po-bya, and the sMin-gling rNgu-ra.

About the twelfth century, a descendant of the leading clan of A-lcags 'Bru, 'Bru lHa-rgyal, being on bad terms with Ge-sar of Gling, left Gling for 'Gu-log. There he and his son A-'bum defeated the three branches of mKhar-ba tribes, who were living in the Bayankara mountains, and took over all their territory. A-'bum's wife came from the Rlangs clan, which, as we have seen, was connected to Sum-pa as far back as the era of the twelve little kingdoms. His son 'Bum-g·yag is said to have married a daughter of the mountain god gNyan-po-g·yu-rtse, and their son Phag-thar established castles at rDo-khog, 'Dzi-khog, and sMar-khog. These three lands, together with rTsang-khog, rNga-khog, and gYu-khog, are known as rMa-shod-drug. Phag-thar's son became a monk at Kaḥ-thog, following the example set by 'Bru lHa-rgyal. He had been a student of rDo-rje-rgyal-mtshan, also known as gTsang-ston, a disciple of Kaḥ-dam-pa bDe-gshegs (1122–1192).

THE DMU TRIBE

The dMu tribe is led by the clan of dMu-tsha rKa/sGa and is famous for its great learnedness. The Po-ti-bse-ru traces the dMu ancestor back to the dMu wife of A-mi Mu-zi-khri-do, while other sources give the first dMu ancestor as sKyin-pa-thang-rje, a descendant of the srid-pa'i-lha. dMu is closely associated with the

Bon tradition and with Zhang-zhung. dMu is also a type of nonhuman sky divinity, and folk traditions connect the first king of Tibet with the sky gods dMu and Phya. A number of ancient kings married queens from the dMu gods. dMu may also be related to the Tibetan word for human, Mi, and to Mu, the name of a particular kind of manlike, red-faced monkey that some of the Ch'iang tribes believed was their ancestor.

But the lineage of the dMu tribe is not well known. Bon texts such as the Srid-pa'i-rgyud-kyi-kha-byang-chen-mo contain a substantial amount of information on the dMu and the related clan of Phya, as well as details on Zhang-zhung culture and language and the Bon tradition. Study of this and other Bon texts will help develop a better understanding of Zhang-zhung culture, and may contribute to research on the tribes as well. Bon texts contain accounts about bSe, Khyung-po, 'Bru, and other tribes worthy of study, but such texts are not always very clear. The language is sometimes obscure and materials of all kinds are mixed together without much organization. Valuable old records may be combined with accounts that are difficult to verify.

TRIBES IN ANCIENT CHINESE RECORDS

References to tribes living across the Tibetan plateau in the times before the Dharma Kings can be found in ancient Chinese records. How these tribes might be connected to the four great tribes of Tibet remains an open question. Western scholars see similarities between the four cultures ('A-zha, Sum-pa, Zhang-zhung, and Mi-nyag) and the descriptions of tribes mentioned in some of the old Chinese annals.

R. A. Stein has pointed out that a particular Ch'iang tribe, the Tang-hsiang, which controlled much of the eastern borderlands by the 5th century A.D., split into two groups in the seventh century. T'ang dynasty records show that one group moved to the south and the other to the north, perhaps explaining the existence of two Mi-nyags. The Tang-hsiang in the north are said by

Chinese annalists to be connected with the ruling dynasty of the later Hsi-hsia state. Mi-nyag is connected by Tibetan historians to the lDong tribe; whether lDong or bSe might be associated with the Tang-hsiang tribe is not yet clear.

Stein, Demiéville, Thomas, and other scholars propose that Sum-pa is connected with the Ch'iang tribes called Su-p'i mentioned in records of the T'ang dynasty (618–907). The Su-p'i were closely associated with a kingdom known in Sui dynasty (581–617) accounts as the Land of Women, located in the rGyal-mo-rong area. Many of the leaders of the Land of Women had the name Tong or T'ang, which Stein believes is Chinese for sTong. The Su-p'i seem to have been located around Nag-chu-kha in T'ang times, and were known as the largest tribe on the plateau. They may have spread all the way to Khotan. Another Kingdom of Women was reported just south of Khotan. Texts relating the history of Khotan, such as the Ri-glang-ru-lung-bstan-pa and the Dri-ma-med-pa'i-'od-kyi-zhus-pa, note Sum-pa tribes near Li-yul (Khotan). Thomas pointed out that third or fourth century Kharoṣṭhī documents from Central Asia mention tribes called Supiya making raids on Central Asian cities, and perhaps they are connected with the Su-p'i or Sum-pa.

OTHER PEOPLES ON THE PLATEAU

Another people in Central Asia that may have had connections with the ancient Tibetan tribes are the Yüeh-chih, a people thought to have spoken an Indo-European language, who lived just north of Koko Nor in the second century B.C. Han dynasty records show that in 175 B.C. some Yüeh-chih moved west past Khotan, eventually becoming part of the Kuṣāṇa Empire, while other Yüeh-chih joined the Ch'iang tribes. The Khotanese scholar Bailey has suggested that the 'Gar clan might be related to the Yüeh-chih, but Tibetan genealogies present a different picture.

There is also scholarly interest in the ancient history of the far western edges of the Tibetan plateau. The west may have been inhabited by Indo-European Dardic peoples before 1000 B.C., and remnants of Dardic peoples are said to still live in Ladakh and Swat. Languages such as Kashmiri and Shina, spoken in some of these regions, are classified by modern linguists as Dardic. There are interesting references to people known as Darada, Derdai, or Dardae along the Indus river near Gilgit (Bru-sha) in early Greek and Roman sources such as Ptolemy (second century A.D.), Strabo (first century A.D.), and Pliny (first century A.D.).

Whether these ancient Dardic cultures were associated with Zhang-zhung, which is thought to have extended to the north and the west in ancient times, remains to be shown. Some modern scholars have considered the possibility that the Zhang-zhung language was Indo-European, while others consider it closely related to Tibetan. Zhang-zhung and Bon are often associated with the "Western regions" and with sTag-gzigs (Persia) in the Bon tradition. A number of Tibetan histories mention that Bon practices from lands to the west entered Tibet in the time before the Dharma Kings, from regions between India and Persia such as Bru-sha. Modern scholars have been interested in investigating possible influences on early Bon from Mesopotamia, Persia, Central Asia, India, and China, as well as comparing Bon with the shamanist practices of ancient peoples around the world.

Tibetologists also see connections between Zhang-zhung and the Yang-t'ung, a Ch'iang tribe reported by T'ang historians to be living in western Tibet in the seventh century. They practiced elaborate burials, animal sacrifices, and divination, practices that might be related to some kinds of Bon. Whether these tribes might be connected to earlier populations in the northwest or with the dMu tribe is not known.

The culture of 'A-zha has also attracted the attention of Western scholars. Pelliot found the name of 'A-zha in old Chinese accounts about the T'u-yü-hun, which supported the idea that these tribes were the same people as the 'A-zha. The T'u-yü-hun were reported by Chinese annalists to be descended from tribes north of China identified by scholars as Turkish-Mongol nomads. They entered the Koko Nor region in the fourth century A.D., establishing a successful kingdom that expanded across Central

Asia to Lop Nor, and even raiding Khotan. F. W. Thomas thought that the 'A-zha might be a Ch'iang culture that had fallen under the domination of the T'u-yü-hun.

It is interesting that when the Chinese pilgrim Sung-yin traveled through the 'A-zha region in 518 A.D., he noted that 'A-zha had a written language; this was a century before the minister Thon-mi Saṁbhoṭa devised writing for spoken Tibetan. Annals from 'A-zha have been discovered at Tun-huang, but they are written in Tibetan, which had become the administrative language of this region by the time of the Dharma Kings.

When the relationships among the cultures of 'A-zha, Zhang-zhung, Sum-pa, and Mi-nyag and the original tribes of Tibet are better understood, it may be possible to know whether the tribes mentioned in foreign records are connected with the tribes in the Tibetan records. Eventually, scholars may be able to trace any threads that might link these various peoples.

REGIONAL NAMES IN LATER TIMES

Over many centuries, the Tibetan tribes spread across the plateau, inhabiting the valleys and mountainsides everywhere but in the driest reaches of Byang-thang. The map above shows approximate locations for some of the different districts that developed over time. Lack of space makes it difficult to represent accurately the extent of various regions, and also makes it necessary to leave out many important place-names. Tracing the founders and development of these and other districts would be one way to help fill in some of the gaps in the history of the Tibetan tribes. Local chronicles and genealogies might prove useful sources for such a study.

RECORDS OF ANCIENT TRIBES

Records from the Han, Sui, and T'ang dynasties mention numerous tribes living on the Tibetan plateau in the period before the 7th century when the kings in sPu-de's lineage from the Yar-lung valley conquered all of the peoples on the plateau. The relationships among tribes named in the Han, Sui, and T'ang annals and lDong Mi-nyag, dMu Zhang-zhung, sTong Sum-pa and bSe 'A-zha, are not yet clear. These four tribes are said to be the ancestors of the Tibetan people.

THE 'A-ZHA KINGDOM

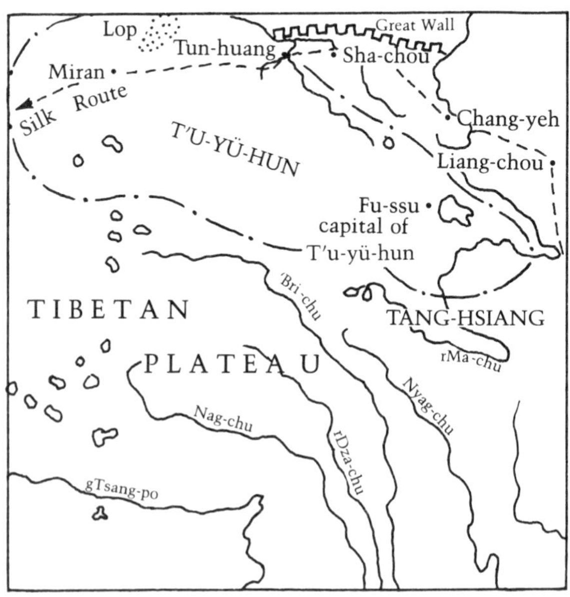

In the northeast corner of the plateau, the T'u-yü-hun, Turko-Mongol peoples from northeast Asia, established a kingdom in the 4th century, a time when northern China was also governed by Turko-Mongol dynasties and tribes perhaps related to Tibetans. At their height, the T'u-yü-hun held some kingdoms along the Silk Route and raided Khotan. They were conquered by Tibet in the 7th century. The T'u-yü-hun seem to be the people known as 'A-zha in Tibet.

CENTRAL TIBET

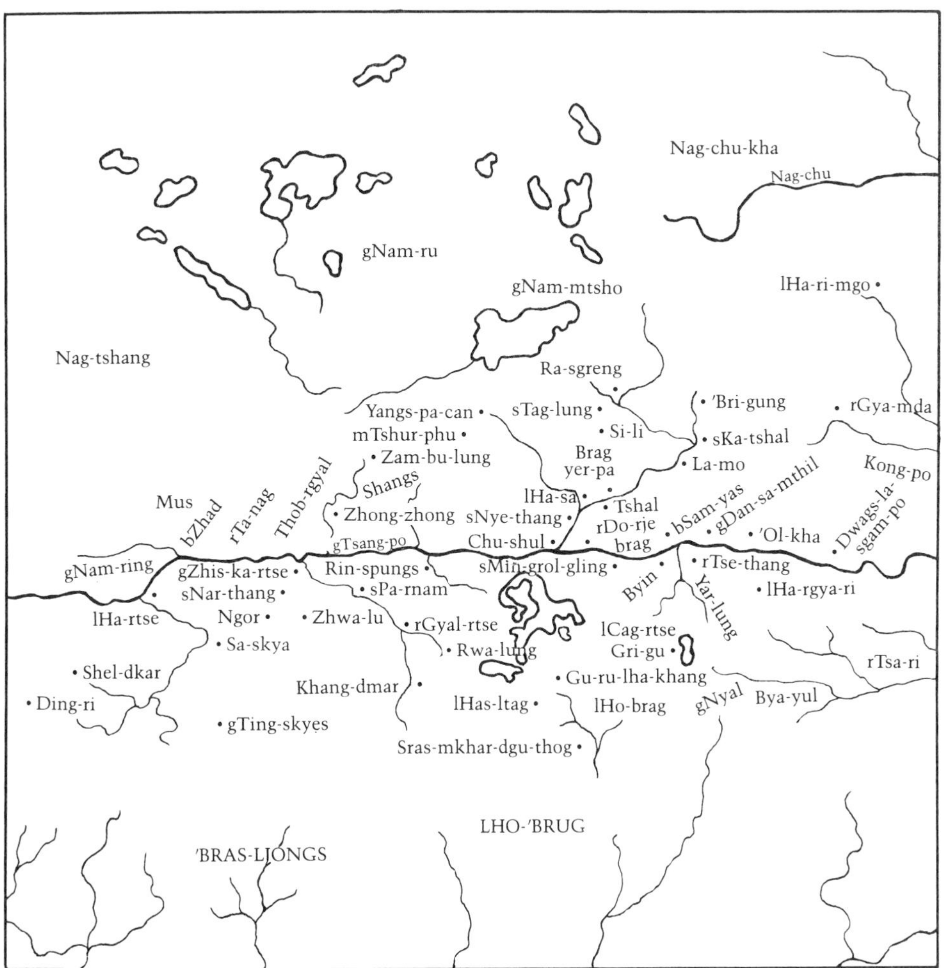

According to most Tibetan histories, Yar-lung in central Tibet was the original home of the Tibetan people and was the center of the culture for hundreds of years. lHa-sa, just to the north, became the capital in the seventh century. Over the succeeding centuries, numerous towns, estates, and religious centers were established along the gTsang-po river and in the valleys of its tributaries. Many more place-names could be added to this map, which, for lack of space, shows only a few of the famous sites in central Tibet.

CHAPTER NINE

THE FIRST TIBETAN KINGS

The lineage of Tibet's kings, according to both Buddhist and Bon accounts, is separate from the lineage of the tribes. The kings are said to have come from outside Tibet, either from the heaven realms or from a royal dynasty in India. To some minds, there is no contradiction between these alternatives, for the kings were said to be extraordinary godlike individuals.

The Bon tradition states that the first king came from a lineage of sky gods, and that he descended on a mountain in Tibet via a dMu rope that linked him to the heavens. Marrying a queen from among the nonhumans, this first king had a son who became king at the proper age. Using the sky rope, the father then returned to the heavens, disappearing like a rainbow. These god-kings continued to arise in Tibet for seven generations. In the eighth generation, the sky rope was cut, and the kings began to leave their bodies on the earth when they died.

Although the Buddhist historians considered the first king a Bodhisattva who appeared in Tibet to prepare the way for the Dharma, they agreed with the Bon tradition about the unusual powers of the early kings. But they looked to India for his ancestors. It is curious that an Indian prince would be thought capable of ascending to the sky, for this trait is not usually ascribed to Indian dynasties. Such questions did not hold much interest for early historians, who were more concerned with determining the particular dynasty to which this first king belonged, and connecting Tibet's royal lineage with India.

ANCIENT INDIAN DYNASTIES

Historians such as Bu-ston, 'Gos Lo-tsā-ba, and many others base their discussions on the account of the eighth century translator Shes-rab-go-cha (Prajñāvarman), who describes the flight of an Indian lord Rūpati during the Pāṇḍava wars between the Pāṇḍu brothers (sKyabs-seng) and the Kuru family. Disguised as women, the prince and his army crossed the Himalayas, and their descendants are said to be the Tibetans.

Many arguments have been made concerning the identity of this Rūpati. The Deb-ther-dmar-po mentions the Mountain Śākyas, the Great Śākyas, and the Śākya Licchavis as possible lineages, and notes that the Ka-khol-ma favors the Mountain Śākyas. The Blue Annals decides in favor of the Licchavis by relying upon predictions in the Mañjuśrīmūlatantra that the Tibetan kings will arise from the Licchavi lineage. In the time of the Buddha, the Licchavis were part of the Vṛjji confederacy with its capital at Vaiśālī. Although they were eventually conquered by Bimbisāra's successor Ajātaśatru, their lineage endured many centuries, for the fourth century A.D. king Candragupta married a Licchavi princess.

To establish the identity of Rūpati, historians considered a number of kings contemporary with the Buddha, kings who also seemed connected with the Licchavis and with the original king of India, Mahāsammata (Mang-po-skur-ba): gSal-rgyal (Prasenajit) of Kosala; King gZugs-can-snying-po (Bimbisāra) of Magadha; King Shar-ba or 'Char-byed (Udayana) of Vatsa or Kośāmbī; and King gTum-po-rab-snang (Caṇḍapradyota) of Avantī or Ujjainī.

One account notes that King Shar-ba's son was born with unusual marks, webbed fingers and eyelids closing from below, and so was cast into the Ganges in a box. Rescued by peasants, the child grew up and ran off in grief to the Himalayas, where he became the ruler of Tibet. Some sources tell such a story about the fifth son of Bimbisāra's youngest son, while other theories say the Tibetan king was the fifth son of Prasenajit or the son of Śuddhodana, father of Śākyamuni Buddha; some favor the lineage of Śatānīka (dMag-brgyad-pa), father of King Shar-ba of Vatsa.

Historians defended favorite theories and disputed those of other authors, presenting not only alternative lineages for the first king, but also various dates for the Buddha, different lists of kings mentioned in the Buddhist canon, and different stories about the Pāṇḍava wars. All these issues were linked together in individually coherent, but mutually conflicting chronologies.

For example, one author cites texts that seemed to place the Pāṇḍavas at the time of the Buddha, while another follows a tradition that placed these wars at the beginning of the Kalī yuga, thus fixing the date for Rūpati in an era long before the lifetime of the Buddha Śākyamuni. The time of the Pāṇḍava war, which is recounted in the Mahābhārata, is still unsettled, and scholars offer dates ranging anywhere from about 3000 B.C. to 800 B.C.

However these issues are decided, we can accept the account of Rūpati, for which there are presumably good sources. It seems possible that descendants of an Indian prince and his followers could have settled in Tibet, perhaps becoming important leaders. But if we were to take Rūpati's soldiers to be the ancestors of the Tibetan people then we would have to ignore the Maṇi bka'-'bum tradition about the original tribes. We would also be unable to account for the individuals who met the newcomer or for the little kingdoms, who agreed to make him their ruler.

ARRIVAL IN YAR-LUNG

When the first king arrived in Tibet, he descended upon a mountaintop, lHa-ri-gyang-mtho in mChims Kong-po, and from there proceeded to lHa-ri-yol-ba or Rol-pa-rtse near the confluence of the gTsang-po and Yar-lung rivers. There he saw the

Yar-lung valley with its deep blue river flowing down from Yar-lha-sham-po peak. The Tun-huang chronicles relate how the first king, gNya'-khri-btsan-po, a god of the sky, descended to earth to be king in Tibet:

"A beautiful country in the center of the world,
a land of high mountains, a pure land, an excellent land . . .
Like rain falling from heaven to nourish the earth,
gNya'-khri came to Tibet
to be the first of the fathers of the country."

gNya'-khri then traveled up the valley to a place called bTsan-thang-sgo-bzhi where he was greeted by a group of Tibetans. When the people first met gNya'-khri, his power and charisma apparently impressed them as superior to that of their own tribal leaders. According to all accounts he was remarkably beautiful but strangely different from other men, and people certainly wondered where he came from. Not speaking the local language, gNya'-khri explained his presence by simply pointing to the sky with his finger. His precise gesture is not described, but the people interpreted it to mean he had come to them from heaven.

It seems that the group who met gNya'-khri attached special importance to the sky, and they may have assumed that their unusual guest had arrived purposely for the sake of their people. They must have had some reason to have gathered together in Yar-lung, and the combination of circumstances must have impressed them as a dream come true. Rejoicing in their good fortune, they raised him upon their necks and shoulders and carried him off to be their ruler. And thus he became known as the Neck-Enthroned King, gNya'-khri-btsan-po.

A GODLIKE KING

We know little else about this mysterious encounter, but judging from later histories, the ancient poetry from Tun-huang, and the inscriptions from the days of the Dharma Kings, the people had remembered and venerated this first king for centuries. He was considered a benefactor and a god, perhaps not unlike the immortal gods of the Greeks, who resided on Mount Olympus and yet took human form at will.

The pillar at Khri-srong-lde-btsan's tomb (eighth century), for example, describes the first king as a son of the gods, the one who established good customs for the people, and whose rule was unchanging. Similarly, the ninth century treaty between Tibet and China contains a passage on the first king, who is referred to as a ruler with great wisdom and understanding. His knowledge of the arts of war protected the people, and his knowledge of the proper customs brought harmony to the land. T'ang historians knew of the first Tibetan king, whose name they presumably learned from Tibetan foreign ministers visiting their court beginning in the seventh century. It is rather amazing that even Muslim historians such as Gardīzī, writing in the eleventh century, had heard of how the first Tibetan king had come from the abode of the gods in the sky.

Old records clearly say that the story of gNya'-khri is not just a myth; that though he was magical, he did actually arrive in Yar-lung and become the king. Other civilizations from this ancient era have also left us records of rulers and heroes who were considered divine—the Egyptian Pharaohs, the Chinese Emperors, the Divine Kings of the Persians. Though these accounts may seem incredible to modern people, it may be that different human capacities manifest in rhythm with each cycle of civilization. Just as we tap powers of nature and perform feats unimaginable to our ancestors, perhaps some of these ancient people had abilities different from ours.

THE ROLE OF THE KING

The bKa'-thang-sde-lnga interprets the purpose of gNya'-khri's arrival in far-reaching terms. Holding a central location in Asia, Tibet was surrounded for centuries by other ancient and powerful civilizations: China to the east, India to the south, Persia to the west, and Ge-sar's kingdom to the north. Until Tibet was united, she would be at the mercy of such strong neighbors. This source also offers a revealing social and psychological portrait of the internal situation before the arrival of the first king: The people's hearts were ruled by hatred and fear, for disaster could strike from every side—from poisoning and thievery, from wild animals, from the curses of one's enemies. gNya'-khri had the power to set society running smoothly because he could establish the proper relationships between people, between countries, and between the world of human beings and other realms and powers.

The Bon tradition gives an interesting interpretation of the role of the first king, describing how he underwent trials and tests of magical skill. Doing battle with demons, he used special weapons, such as armor that places itself on the warrior and a spear that hurls itself at the enemy. Upon his arrival, gNya'-khri subjugated the local shaman, O-yong-rgyal-ba from Sum-pa, and defeated the ruler of gNubs, one of the twelve little kingdoms. Buddhist historians agree that these early kings had magical powers, that they could fly through the sky and penetrate rock.

FOUNDING THE NEW DYNASTY

Near the bank of the Yar-lung river, the new king built Yum-bu-bla-sgang castle, which could still be seen until very recently; he took gNam-mug-mug as his queen, and established the kingdom. Some of the oldest accounts say that Yum-bu castle was built by a later king, lHa-tho-tho-ri, while gNya'-khri's castle was called Sham-po-dgu-brtsegs. In any case, he made the Yar-lung valley his royal seat, and the twelve little kingdoms that had warred among themselves became his subjects.

gNya'-khri and the six generations after him are known together as the Seven Thrones, Khri-bdun. When the king's first son was old enough to ride a horse, the father would hand over the kingdom and disappear into the sky, leaving no corpse behind upon the earth; thus the histories speak of these kings as having had the sky for their tombs.

The names of the ancient Tibetan kings are given in many sources. The lists generally agree, differing in spelling and in grouping of kings. The son of gNya'-khri and gNam-mug-mug was Prince Mu-khri, who married Sa-ding-ding. Their son was Prince Ding-khri, who married So-tham-tham and had a son named Prince So-khri, who married Dog-mer-mer. Their son was Prince Mer-khri, whose wife was gDags-kyi-lha-mo-dkar-mo. Their son, Prince gDags-khri, married Srib-kyi-lha-mo; their son, named Prince Srib-khri, married Sa-btsun Rlung-rje.

THE KING'S LINEAGE AND THE PEOPLE'S LINEAGE

The lineage of rulers founded by the first king is said to have remained completely separate from the tribes of the Tibetan people for a long time. Most traditional histories point out that the kings did not begin marrying among the Tibetans until ten generations or so before the seventh century king Srong-btsan-sgam-po. These kings instead took wives from among the gods and nature spirits — lha, klu, dmu, and gnyan.

Yet we have seen in chapter 8 that some sources connect the Tibetan tribes to the gods, and the god-lineage or lha-rigs of the tribes is especially famous for its connections with non-human realms. Perhaps some of the ancient queens came from the god-lineage of the tribes. The opening lines of a list of little kingdoms from Tun-huang suggest that ancient marriage alliances were made with women from the regions of Yar-lung, mChims, Dags-po, and sKyid.

THE ERA OF GNYA'-KHRI BTSAN-PO

Tibetan historians have preserved a number of possible dates for gNya'-khri-btsan-po. One well-respected tradition states that 660 years elapsed between the first king and a later king, lHa-tho-tho-ri; another 150 years elapsed between lHa-tho-tho-ri and King Srong-btsan-sgam-po. lHa-tho-tho-ri lived 120 years, and these must be added to the calculations. The exact date for gNya'-khri would depend upon the date used for Srong-btsan-sgam-po, who lived in the seventh century. But this gives a rough date for the first king in the fourth century B.C.

A slightly different tradition suggests 500 years between gNya'-khri and lHa-tho-tho-ri, 120 years for lHa-tho-tho-ri, and 111 years to Srong-btsan-sgam-po. This would put the first king in the second century B.C.

Judging from the number of kings in the Tibetan dynasty, we would expect forty-two generations or approximately 1050 years to elapse between the first king and the last, who is known to have died about 841 A.D. This would indicate a date in the third century B.C.

Thus a date somewhere between the fourth and second century B.C. appears most reasonable. On this basis we can rule out various possibilities offered by our sources, such as 447 B.C. 794 B.C., or a date some 2000 years after the Buddha's nirvana.

Other historians give more likely dates between the fourth and second centuries: a wood-tiger year, either 247 B.C. or 307 B.C., with 247 B.C. perhaps the more likely (see chapter 11). We cannot be sure whether this wood-tiger year marks his birth or his arrival in Tibet. Although either of these dates makes it more difficult to connect gNya'-khri to the sons or grandsons of kings in the time of the Buddha, he may have been one of their descendants.

The T'ang annals mention an account of the origin of the first Tibetan king that would give a very different date for his arrival. A prince from the Liang dynasty, which ruled northern China in the fourth and fifth centuries A.D., is said to have fled China in 414 A.D. and become the ruler of the T'u-fan (Tibetan) tribes. But the T'ang historians do not especially credit this story, and it does not seem particularly relevant to gNya'-khri. Members of some of these fourth and fifth century dynasties, many of which were established by Turkish and Mongol peoples, may well have joined the tribes in Tibet at various times, but a fifth century individual could not be gNya'-khri btsan-po!

The era of gNya'-khri-btsan-po's unification of the twelve little kingdoms seems to coincide with the unification of the warring states in China under the leadership of the short-lived Ch'in dynasty (221–206 B.C.); they were succeeded by the Han dynasty (202 B.C.–220 A.D.). Likewise in Greece, rival city-states were finally united by Philip II of Macedonia, who was followed by his famous son Alexander the Great (356–323 B.C.), conqueror of much of western Asia. A contemporary of Alexander's was Candragupta, who founded the Mauryan Empire in 322 B.C., uniting the Indian kingdoms. His grandson, Dharma King Aśoka, ruled almost the entire Indian subcontinent.

If we consider the third century B.C. as the beginning of the organization of the kingdom of Tibet, then nine hundred years passed till the days of Srong-btsan-sgam-po, fifteen hundred years passed till the days of Sa-skya Paṇḍita, and over twenty-three hundred years passed until the end of the traditional government of Tibet some thirty years ago. Among all the older civilizations that have continued into modern times, Tibetan civilization may be one of the oldest.

gNya'-khri-btsan-po

THE ERA OF GNYA'-KHRI-BTSAN-PO

In the middle of the third century B.C., Central Asian states of Khotan and Kucha were established, and the great Dharma King Aśoka ruled India. The period of the Warring States in China was finally ended when the Ch'in dynasty united China. In the west, Seleucus, a general of Alexander the Great, had established an extensive empire, but the Parthian Persians soon became independent. Rome now controlled all of Italy, but was not to expand greatly for another century.

ANCIENT INDIAN DYNASTIES

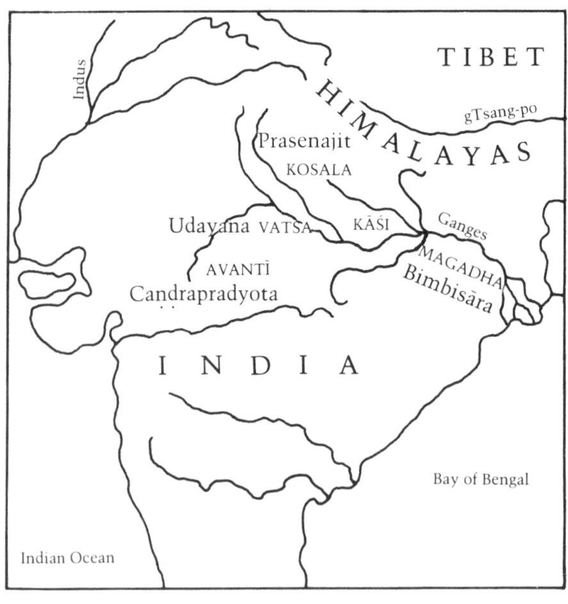

According to numerous Tibetan histories, the first Tibetan king was descended from one of the royal Indian families in power at the time of the Buddha. Northern India in this era was divided into kingdoms known as mahājanapadas, each ruled independently. From one of these old kingdoms, an Indian prince is said to have fled or been cast out, making his way across the Himalayas and into Tibet where he was given the ruling title of gNya'-khri-btsan-po.

THE ERA OF GNYA'-KHRI-BTSAN-PO

TIBET	INDIA	CHINA	WEST
500 B.C. ? Little kingdoms war with each other	881 B.C. A traditional date for Parinirvāṇa of the Buddha	403–221 B.C. Warring States Era: period of disunity	558–530 B.C. Cyrus the Great founds Achaemenid Empire in Persia
247 B.C. gNya'-khri-bstan-po, first king of Tibet, arrives in Yar-lung valley	483 B.C. Parinirvāṇa of the Buddha (Western view)	350 B.C. Taoist, Confucian and Legalist schools develop	490–449 B.C. Persian Wars: Greece defeats Persia
Sum-pa shamans subdued	327 B.C. Alexander the Great of Macedonia invades India	221–206 B.C. Ch'in dynasty	427–347 B.C. Plato
Little kingdoms unite under rule of gNya'-khri; first castle built in Yar-lung	321 B.C. Candragupta establishes Mauryan Empire	215 B.C. Section of Great Wall completed	384–322 B.C. Aristotle
c. 250 – c. 100 B.C. 7 Khri: gNya'-khri Mu-khri Ding-khri So-khri Mer-khri gDags-khri Srib-khri	269 B.C. Aśoka rules Mauryan Empire; converts to Buddhism	202 B.C. Han dynasty unites China till 220 A.D.; early capital at Ch'ang-an	c. 350 B.C. Rise of Macedonia
175 B.C. Little Yüeh-chih tribes join Ch'iang in far northeast	c. 250 Khotan established by Áśoka's son	200 B.C.–200 A.D. Turko-Mongol Hsiung-nu tribes rise to power on China's N. border	307 B.C. Alexandria built in Egypt: center of Greek learning
c. 115 B.C. Some Ch'iang tribes known to be allied with Hsiung-nu in far northeast	c. 185 B.C. Greek king Demetrius rules Bactria; Greek influence extends into N.W. India	200 B.C. Yüeh-chih tribes living in N.W. borderlands	300 B.C. Olmec culture begins to unify Middle America; sun temple built at Teotihuacan
c. 100 B.C. Jo-ch'iang tribes in far north along Kun-luns, near Central Asian trade routes through Khotan	183 B.C. Puṣyamitra Śuṅga usurps power in palace revolt; establishes capital at Videśa; Śuṅga dynasty rules N. India till 72 B.C.	175 B.C. Hsiung-nu push most Yüeh-chih west into Ferghana	300 B.C. Farming cultures in southwestern North America; hunting and farming in eastern North America
	c. 160 Greek king Menander rules Bactria	138–126 B.C. Central Asian expedition of Han general and envoy, Chang-ch'ien	287–212 B.C. Archimedes
		112 B.C. Silk Route opens across Central Asia	280–200 B.C. Rome begins expansion
			147 B.C. Greece comes under Roman control

CHAPTER TEN

THE EARLY TIBETAN CULTURE

The era following the reigns of the Seven Heavenly Thrones opens up a new chapter in Tibetan history. Once the godlike kings began to die upon the earth, many changes took place in Tibetan culture. Although the full significance of these changes is not discussed in most histories, the major events are well known and can be found in many sources.

The son of the last of the Heavenly Throne Kings, Sribs-khri, and his queen Sa-btsun Rlung-rje, was Gri-gum-btsan-po. After his reign began, Gri-gum fell under the influence of one Bon-po 'A-zha, a powerful shaman from Gurṇavatra, a region said to be somewhere between India and Persia. The king came to rely heavily upon his advice and instruction because the shaman made impressive predictions. Gri-gum decided to ask his mentor about the odd name he had received as a child, for Gri-gum means "Slain by a Sword." When the shaman replied that Gri-gum was bound to die violently, the king believed him.

Intent on determining at least a part of his fate, Gri-gum chose to go into battle and challenged each of his ministers, but no one dared accept. Finally he insisted that Lo-ngam, the keeper of the royal horses, engage in a contest with him. Lo-ngam agreed, and devised an elaborate plan to trick the king into relinquishing all of his magical powers.

Playing the loyal servant, Lo-ngam convinced the king he wished him to win and told him how to assure his victory: Sacks of cinders must be loaded on the backs of cattle and sharp tips affixed to their horns; the king should wear a black turban, fasten a mirror to his forehead, and wear a fox corpse on one shoulder and a dog corpse on the other. If he then waved his sword fiercely above his head, he would surely win. Perhaps imagining that he might alter his fate by following these strange instructions, Gri-gum prepared everything as Lo-ngam had instructed.

The contest took place in Myang-ro west of Yar-lung on a specially chosen day. In the first moments of the struggle, the cattle were loosed. Charging at each other, they pierced the bags of cinders with their horns, creating a dense haze of dust. The king found himself bereft of divine support because the corpses on his shoulders had chased away his protector gods. Waving his sword above his head, Gri-gum accidentally cut the rope that connected the king to the realm of the sky, leaving himself completely unaided. In the midst of the dusty confusion, Lo-ngam aimed an arrow at the flashing mirror and killed the king.

EXILE AND RETURN OF THE PRINCES

Gri-gum's corpse was the first left on the earth by a Tibetan king. It was placed in a copper box and thrown into the river where it floated east to Kong-po and came into the possession of a nāga. In the meantime, Gri-gum's sons by Queen Klu-srin-mer-lcam — Bya-khri, Nya-khri, and Sha-khri — had fled to Kong-po, sPo-bo, and Dags-po, flying off on their father's magical white ox. Lo-ngam took the throne and made Gri-gum's queen his wife, forcing her to assume his old job of caring for the horses. But one night the god Yar-lha-sham-po in the form of a shining white man came to the queen in a dream. As she awoke, she saw a magnificent white yak leaving the grounds. Months later, the queen gave birth to a lump of blood that she placed in the horn of a yak, keeping it warm and feeding it milk. Soon a beautiful child emerged, known as Ru-la-skyes, Arisen From A Horn.

As this magical child grew up, he asked about his father and his family, and the queen told him of his exiled uncles and the death of her husband. Ru-la-skyes set out for Kong-po where he recovered the corpse of Gri-gum by searching out an unusual ransom to give the nāga: a girl whose eyes opened from below like those of a bird. When he brought the casket back to Yar-lung, the corpse began to moan. Thus this burial ground is known as Ngar-pa-thang, the Plain of Moaning.

After defeating Lo-ngam and one hundred of Lo-ngam's followers, Ru-la-skyes helped his brothers return to power. Bya-khri became the new king, and one brother became the White Prince of Kong-po. When the queen mother prayed for the success of the new kingdom, the gods announced: This son will be the best. And this became his ruling title—sPu-de-gung-rgyal. The new king built 'Phying-ba-stag-rtse, Tiger Peak castle, in Yar-lung on a mountainside overlooking the 'Phyongs-rgyas river.

The Queen Mother Mer-lcam remained concerned for the future of her son's kingdom, for she was pregnant with the child of Lo-ngam. Afraid the unborn child might grow up to claim the throne, she asked to be buried alive beneath Tiger Peak castle.

ADVANCES IN TIBETAN CULTURE

sPu-de-gung-rgyal thus began his reign at Yar-lung with the sacrifice of his mother and with the aid of the magical Ru-la-skyes. So competent and heroic was this half-brother that the king called him "Uncle" and made him first minister. The king also gave him a new name, lHa-bu-smon-gzung, and from his lineage came six more devoted and magical ministers, who advised the next six kings.

These ministers are credited with many inventions and the advancement of Tibetan culture. In the time of Ru-la-skyes, uses for charcoal were discovered, and techniques for smelting and working gold, silver, copper, and iron were developed. Bridge-building was greatly improved, making travel and trade much easier. Ru-la-skyes especially promoted agriculture, designing yokes for harnessing draught animals and developing more efficient irrigation methods.

The early kings were aided not only by great ministers, but also by shamans who provided the rulers with predictions, divinations, and various protective rituals. That shamanist practices were now of great importance in Tibetan culture we know from the stories about a strange child who arrived in Tibet during this period. A descendant of the old gshen or shaman lineage, he later came to be known as gShen-po-che. He had donkey ears that he covered with a turban, and he roamed about Tibet until he was twelve years old, when the gods and demons carried him off into their realms.

When he returned twelve years later, gShen-po-che had much knowledge about the gods and demons. He could explain why people fell ill and knew how to make the crops grow. Knowing how to interpret signs and how to make the proper offerings to please the powerful local spirits, he performed rituals to bring benefits to the land. During these ceremonies even the common people could hear heavenly music. Possessed of great magical power, he could ride through the sky on a shaman's drum and perform special rites for the dead. The doctrines he preached foreshadowed the Buddhist teachings of Vinaya and Abhidharma, and Buddhist historians interpret these early events as signs of the future success of Buddhism in Tibet.

TOMBS OF THE TIBETAN KINGS

Since Gri-gum had cut the sky rope, sPu-de-gung-rgyal and succeeding kings now left corpses upon the earth. Some texts locate the first two tombs on the upper mountain slopes in the bare rocky outcroppings, though other sources place them near Yar-lung, along the 'Phyongs-rgyas river in the region that became the royal burial ground. Tombs of the seventh to ninth century Dharma Kings are still visible in 'Phyongs-rgyas as raised mounds that covered the subterranean tombs.

The tomb of a Tibetan king was considered sacred ground and was guarded by special ministers of the interior. These guards were assigned to the tomb for the rest of their lives, and were not permitted to interact in any way with other people. Even stray animals entering the burial grounds became the property of the guards, for they could not be returned to their owners. In yearly ceremonies when offerings were brought to the royal ancestors, the guards remained hidden until the visitors had left. They then were allowed to come forth and enjoy the gifts left at the tombs.

For the monarchy, the tombs became important retreat centers and were symbolically connected with the passing of royal power to the next generation. When the crown prince came of age, the father king would retire to his specially built tomb and undergo a ceremonial death that conferred the right to rule upon his son. The old king would then live out his days in strict retreat.

The bKa'-thang-sde-lnga gives an interesting description of the burial rites of the early kings, reminiscent of ancient Egyptian practices. The body of the dead king was first anointed with gold dust and then placed in the center of nine enclosures. A golden image of the king was seated upon a throne, which was surrounded by silver, gold, turquoise, and other treasures. An elaborate ritual guided the spirit of the dead king safely to the other world and provided for his needs along the way.

THE DEVELOPMENT OF BON

The Bon tradition has undergone many stages of development, absorbing influences from foreign lands and later from the Buddhist tradition. Some histories note that Bon practices began during Gri-gum's reign, while others say certain Bon practices were then condemned. According to the Bon text Srid-pa'i-rgyud-kyi-kha-byang-chen-mo, Gri-gum's death was connected with the arrival of heretic Buddhists in Tibet. Other sources tell of the arrival of Bon-pos from 'A-zha or from Zhang-zhung and Bru-sha during Gri-gum's reign, while new Bon teachings are sometimes said to have begun in the time of sPu-de-gung-rgyal.

The later Bon tradition, which may be quite different from early Bon, contains nine divisions of teachings, the first four devoted to rituals and practices for healing, making offerings, burial rites, divination, and subjugation of negative forces. The other divisions deal with moral practices, meditation, and more mystical doctrines. It will require extensive research to be able to determine the earliest forms of Bon, what the Bon-pos from foreign lands brought into Tibet in Gri-gum's and sPu-de-gung-rgyal's time, and the effect on the folklore of the people.

Tun-huang documents offer material on ancient religious ideas and mythology, funeral practices, divination techniques, and various kinds of folklore. Later Bon texts, such as the Kha-byang-chen-mo, contain much old material on gShen-rab-mi-bo, the founder of the Bon tradition. His activities in Tibet, China, sTag-gzigs, India, and other regions are described, and the many different Bon lineages are explained. Cosmological ideas are also presented, together with descriptions of many gods and spirits, and their relationships with human beings. A few Bon texts have recently been translated, offering Western scholars additional material to study.

THE ERA OF THE TWO "STENGS" KINGS

The reigns of Gri-gum and his son sPu-de-gung-rgyal were known together as the period of the Two sTengs of the Upper Regions, sTod-kyi-stengs-gnyis. This period appears to have been a critical point in Tibetan civilization, marked by the first death of the godlike kings, the beginning of royal burial practices, and advances in technology, as well as religious and social changes.

Critical political struggles were also going on among various clans. The Tun-huang documents describe a very complicated plot involving several families who vied with each other after Lo-ngam's defeat. These documents, as well as an eighth century inscription from Kong-po, mention only two sons of Gri-gum, rather than three. Later historians preserve several conflicting accounts about which son fled to which land of exile. Perhaps the various traditions about these kings were influenced by local power struggles long since forgotten, or arose from oral traditions belonging to different regions or tribes.

This era of the Two sTengs probably fell sometime between the first century B.C. and the first century A.D. One modern Tibetan historian suggests that sPu-de was contemporary with the Han Emperor Wu-ti (140–85 B.C.). By the first century B.C., the opening of the Silk Route across Central Asia linked Han dynasty China (202 B.C.–220 A.D.) to India, Persia, and the West, allowing communication and trade all across the Old World for the first time. Northern India and western Central Asia were at this time in the control of the Kuṣāṇa kings, the greatest of whom was the Dharma King Kaniṣka. His support encouraged the spread of Buddhism into Central Asia, which was becoming a melting pot for eastern and western philosophies and technologies. Whether any of these influences penetrated into Tibet in this era is an exciting question for future research.

THE SIX "LEGS" AND THE EIGHT "LDE" KINGS

The kings following Gri-gum and his sons continued to marry mi-ma-yin (nonhuman) queens for another fifteen generations. The next six kings were called the Six Good Ones of the Earth, Sa'i-legs-drug. When they died, they were buried in the middle of the mountain slopes among the foothills or on the high rocky peaks. E-sho-legs, the first Legs king, was the son of sPu-de gung-rgyal and 'O-ma-thang sMan-mtsho. The son of Ru-la-skyes, lHa-bu-mgo-dkar, became his minister, and accomplished the domestication of farm animals and the development of more advanced agricultural techniques.

King E-sho-legs and his queen rMu-lcam Bra-ma-na resided in the castle of sTag-rtse. He was succeeded by five generations of Legs kings. Their son Prince De-sho-legs married Klu-sman Mer-mo and lived in the castle of rGod-rtse. Their son Prince Thi-sho-legs married bTsan-mo Gur-sman and lived in the castle of Yang-rtse. Their son Prince Gong-ru-legs married mTsho-sman

'Brong-ma and lived in the castle of Khri-rtse. Their son Prince 'Brong-gzher-legs married gNam-sman Bu-mo and lived in the castle of rTse-mo-khyung-rgyal. Their son Prince I-sho-legs married rMu-lcam dMar-legs and lived in the castle of Khri-brtsigs-'bum-gdugs.

The next eight kings are grouped together as the Eight lDe. Some sources note that their tombs were in the rivers, but there is little information about them except the names of their queens. Za-nam-zin-lde married mTsho-sman Khri-dkar, and their son was Prince lDe-'phrul-nam-gzhung-btsan. He married Se-gnyan Mang-mo, and their son was Prince Se-snol-gnam-lde. He married Klu-mo Mer-mo, and their son was Prince Se-rnol-po-lde. He married 'O-za sDe-mo-mtsho, and their son was Prince lDe-snol-nam. He married Khri-sman rJe-mo. Their son was Prince lDe-snol-po. He married Se-btsun gNyan-mo, and their son was Prince lDe-rgyal-po. He married sMan-btsun Lug-gong, and their son was Prince lDe-spring-btsan. He married Nye-btsun Mang-ma-rje.

THE FIVE "BTSAN" KINGS

The next five generations were known as the Five bTsan, the Powerful Ones. These kings began the practice of marrying among the Tibetan people. The Ngo-mtshar-rgya-mtsho explains that Tibetan princes were now given the title of bTsan-po, Mighty One, whereas before this time, they were called lHa-sras, Son of the Gods. Unfortunately, there is not much information about the Five bTsan, except for names of queens and ministers.

King To-re-long-btsan was the son of lDe-spring-btsan and Nye-btsun Mang-ma-rje. His ministers were 'Bring-po-rgyal of the Thon-mi clan, Dar-rje of the sTong clan, and sMan-dar-re of the sNubs clan. He married 'O-ma lDe-bza' Khri-btsun-byang-ma. Their son was Prince Khri-btsun-nam. His ministers were gNya'-dor-gtsug of the sNubs clan and Khri-zung-mo-khong of the sBas clan. He married sMan-bza' Khri-dkar and had a son named Prince Khri-sgra-spungs-btsan. His ministers were sGra-rdzi-mun of the mGar clan and Glu-ma-dred-po of the Thon-mi clan. His queen was sMan-gsum Klu-steng, and their son was Prince Khri-thog-rje-thog-btsan. His ministers were gNya'-btsan-ldem-bu of the mGar clan and Thog-rje-mong-btsan of the sNubs clan. His queen was Ru-yong-bza' sTong-rgyal-mo-mtsho, and their son was Prince lHa-tho-tho-ri gnyan-btsan, who took as his wife the noblewoman rNo-bza' Mang-dgar.

There were many great ministers during these early times, though it is not always certain under which kings they held office. The Tun-huang documents extol them as unusual and powerful men. They had "sharp eyes" and "knew men's hearts." It is said that one had such wide-ranging intelligence that even when far away, he was as though present. Some were famous for their bravery or their careful advice, and a few were said to be clairvoyant.

THE ERA OF THE "LEGS," THE "LDE," AND THE "BTSAN"

The Six Legs, the Eight lDe, and the Five bTsan seem to have ruled Tibet between the first and fifth centuries A.D. In the same period, the Han dynasty fell and north China was invaded by many foreign tribes. The Huns spread across Central Asia, reaching India, as well as Europe, where the Roman Empire, too, was threatened by wave after wave of barbarians from the steppes. Despite these widespread upheavals, the cities of Central Asia, such as Kucha and Khotan, were becoming important Buddhist centers, inspired by the Buddhist civilization of India, where the disciples of Nāgārjuna were eloquently supporting the Dharma. As Central Asian monks made their way east, the first translations of Buddhist texts took place in China.

The Buddhist tradition had not yet entered Tibet, but through the reign of Khri-thogs-rje-thogs, lHa-tho-tho-ri's father, Tibetan culture had been nourished by the activities of the Bon-pos known as gshen, and by bards, who recounted tales known as sgrung, which were stories of gods, heroes, and great clans. The people met together to hear these stories and to engage in games, debates, tests of skill, and contests where singers known as lde'u performed riddling songs. Some of these recitations seem also to have served as part of religious rituals or played a role in the ceremonies surrounding kingship. Important events were preserved in song or verse, providing the basis for the oral transmission of history. The traditions of the gshen, sgrung, and lde'u gave the people guidance and developed into an extensive folklore that informed the Tibetan way of life for centuries.

Though it would be hard at this stage of research to reconstruct this era, Tibetan culture was certainly growing richer and more complex. As agricultural advances inspired by the great ministers spread among the people, villages expanded. Some records mention that better houses were built using brick and stone topped by wooden roofs. Imposing castles and fortresses were built on defensible hillsides overlooking the valley—perhaps by wealthy nobles—but the social structure of this early society is still mostly conjecture. The ancient-looking stone fortresses in the eastern borderlands, in Kong-po, and in sPo-bo may be examples of this old architecture.

PREBUDDHIST TIBET

Religious and spiritual influences came not only through the Bon traditions, but from the very land itself. The rugged mountains and deep valleys displayed dramatic forms and shapes that served as natural symbols of a higher order, evoking the power of the cosmos.

The Buddhist tradition notes that certain locations were the home of Bodhisattvas and Arhats—hidden peaks such as rTsa-ri; the majestic mountain range gNyan-chen-thang-lha in central Tibet; the sacred lake Manasarowar (Ma-pham) near holy Mount Kailāśa (Ti-se). For untold centuries, pilgrims from India and the northwest had visited Ti-se and Ma-pham. Mantras and handprints of ḍākinīs can be seen naturally impressed in the rock around this lake, and the water has curative properties. White peaks in the Himalayas such as Jo-mo-gangs-dkar (Everest) and Gangs-chen-mdzod-lnga (Kanchenjunga) were recognized as natural power centers, as were mountains to the north such as A-mnyes-rma-chen and gNyan-po-g·yu-rtse. Over the centuries, long before Buddhist teachings reached Tibet, a way of life developed that was closely attuned to the land and its natural powers.

This ancient culture had been presided over by twenty-seven generations of kings up to the last bTsan king. These kings had ruled for six hundred and twenty years since the arrival of gNya'-khri-btsan-po. With lHa-tho-tho-ri, the last of the bTsan kings, a new cultural era was to begin, one in which the Buddhist teachings slowly began to influence Tibetan civilization.

TRADE ROUTES ACROSS ANCIENT ASIA

In 112 B.C. the Silk Route was opened across Central Asia. This trade route linked China and India in the east with Persia and Rome in the west. Goods flowed in both directions along difficult caravan tracks, establishing trade and communication all across the Old World for the first time. But raiding nomadic tribes such as the Hsiung-nu made travel dangerous and threatened the new prosperity of kingdoms along the route, such as Khotan and Kucha. Han dynasty annals report that trade marts were also established in the northeast at Koko Nor.

ANCIENT CULTURAL CENTERS

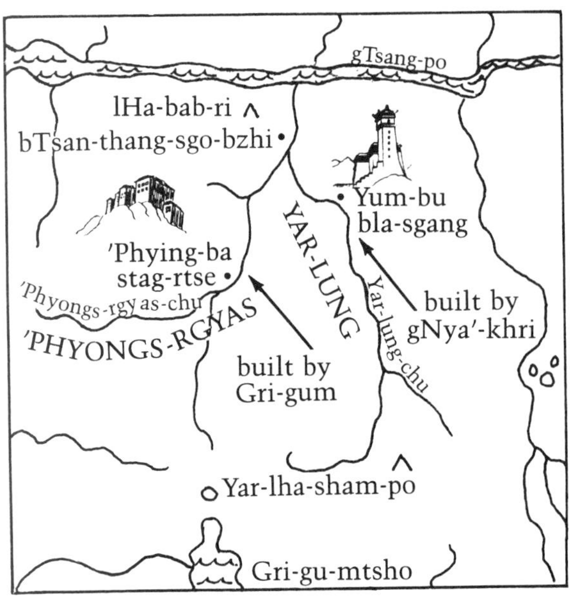

The valleys of the Yar-lung and 'Phyongs-rgyas rivers became the center of the ancient Tibetan civilization. The first godlike king descended upon a mountain in Kong-po and made his way south to lHa-bab-ri. From there, the new king went to bTsan-thang-sgo-bzhi where he first met the Tibetan people. His castle of Yum-bu-bla-sgang and the castle of the eighth king at 'Phying-ba are located here. Only in the 7th century was the capital shifted north to lHa-sa.

EXILE AND RETURN TO YAR-LUNG

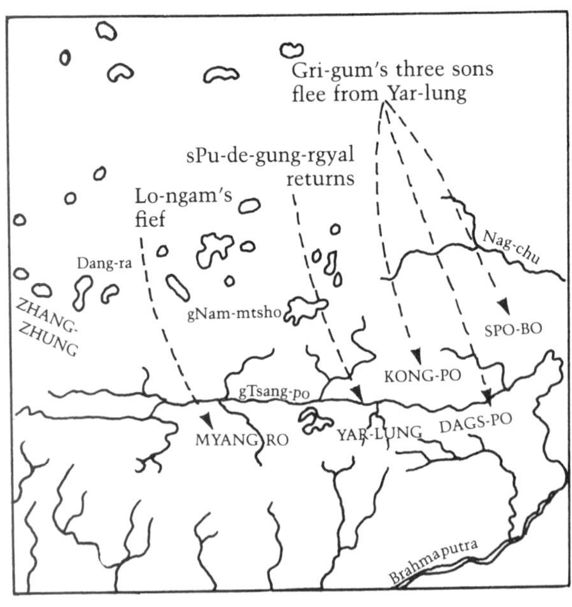

The ancient principalities along the gTsang-po seem to have challenged the newly emerging kingdom established at Yar-lung. After the death of Gri-gum-btsan-po, the eighth king of Tibet, his three sons fled into exile to Kong-po, sPo-bo, and Dags-po while the usurper Lo-ngam ruled at Yar-lung. One son defeated Lo-ngam and returned to the Yar-lung valley where he restored the dynasty as King sPu-de-gung-rgyal. Another son became the White Prince of Kong-po.

THE ERA OF GRI-GUM-BTSAN-PO

TIBET	CHINA	ASIA	WEST
c. 150 B.C. Period of the Two sTengs:	127 B.C. China garrisons Tun-huang, Kan-chou, Liang-chou	130 B.C. Yüeh-chih reach Bactria; push Śaka tribes into India	100–44 B.C. Julius Caesar, Emperor of Rome
Gri-gum-btsan-po rules in Yar-lung	121 B.C. Hsiung-nu driven north of Gobi desert	c. 100 B.C. Buddhist cave temples built at Ajantā in central India	51 B.C. Gaul becomes Roman province
Gri-gum killed by Lo-ngam; his sons Bya-khri, Nya-khri, and Sha-khri flee to Kong-po, sPo-bo, and Myang-ro	112–115 B.C. Ch'iang tribes invade western China	100 B.C.–100 A.D. Rise of Kuṣāṇa Empire, cosmopolitan influence across N. India, Punjab, W. Central Asia; support for Buddhism	31 B.C. Egypt becomes Roman province
Bya-khri returns to power with aid of Ru-la-skyes; takes name of sPu-de-gung-rgyal	112 B.C. Silk Route opens across Central Asia	Khotan and Kucha become Buddhist centers; Central Asian Buddhists travel to China	30 A.D. Jesus of Nazareth is crucified
First burial of kings	9 A.D.–23 A.D. Wang Mang, former regent, establishes Hsin dynasty	50 B.C.–250 A.D. Śātavāhanas rule S. India	45 A.D. St. Paul begins missionary travels
c. 50 B.C. Six Sa'i-legs: E-sho-legs De-sho-legs Thi-sho-legs Gong-ru-legs 'Brong-gzher-legs I-sho-legs	25 A.D.–220 A.D. Later Han dynasty established; capital at Lo-yang	120 A.D.–395 A.D. Central Asian Śakas establish satrapy in W. India	70 A.D. Rome destroys Temple of Jerusalem
c. 100 A.D. Eight lDe: Za-nam-zin-lde lDe-'phrul nam-gzhung-btsan Se-snol-gnam-lde Se-rnol-po-lde lDe-snol-nam lDe-snol-po lDe-rgyal-po lDe-spring-btsan	d. 87 A.D. Ssu-ma Ch'ien writes historical memoirs; model for later dynastic annals	c. 200 A.D. Indian epic poem Bhagavad Gīta takes final form	117 A.D. Roman Empire at its greatest extent
c. 300 A.D. Five bTsan To-re-long-btsan Khri-btsun-nam Khri-sgra-spungs-btsan Khri-thog-rje-thog lHa-tho-tho-ri	148 A.D. Translation center established at Lo-yang by Central Asian Buddhists	225 A.D. Sassanian kingdom established in Persia; extends N.W. to India by 244 A.D.	220 A.D. Goths invade Asia Minor and the Balkans
	184 A.D. Yellow Turban Rebellion: inspired by Taoist messianic movements		250 A.D. Classic Mayan civilization arises
			Hopewell Indian culture by Great Lakes, N. America
			268 A.D. Goths sack Athens, Sparta, and Corinth
			285 A.D. Roman Empire partitioned into East and West

CHAPTER ELEVEN

THE HAPPY GENERATIONS

In the fifth century A.D. influences from the Buddhist tradition that would later revolutionize Tibetan civilization began to reach the Land of Snow. The first king to be touched by the Dharma was lHa-tho-tho-ri gnyan-btsan, who is revered by the Buddhist tradition as an incarnation of the Bodhisattva Samantabhadra. Tibetan historians sometimes call lHa-tho-tho-ri and the kings that succeeded him the Happy Generations, contrasting their era with previous times when the Buddhist tradition was still unknown in Tibet.

Some historians connect lHa-tho-tho-ri with the prediction in the Śrīmālādevī-sūtra that the Dharma would reach the land of the red-faced men 2500 years after the Parinirvāṇa of the Buddha, for "red-faced" is a common epithet of the Tibetan people.

But other histories offer more specific dates for this king; for example a water-snake year of 213 A.D., or a water-ox year of 173 A.D. But second or third century dates for lHa-tho-tho-ri would put over four hundred years between his reign and the reign of Srong-btsan-sgam-po, who is said by most sources to have been only five generations after lHa-tho-tho-ri.

These early dates for lHa-tho-tho-ri also contradict the tradition that places this king approximately 150 years before King

Srong-btsan-sgam-po and 660 years after gNya'-khri. Knowing that Srong-btsan-sgam-po lived in the beginning of the seventh century, we can place lHa-tho-tho-ri's death in the mid-fifth century. Since historians generally agree that lHa-tho-tho-ri lived 120 years, he would have been born in the early fourth century. This seems to be the earliest he could have lived, if we assume the 150 years marks the end of his reign.

This calculation accords with the dates given in some sources: a wood-dog year of either 314 A.D. or 374 A.D. With only five generations between the two kings, it seems that 374 A.D. (1611 years ago) might be the best birthdate for lHa-tho-tho-ri, making 247 B.C. (2233 years ago) the better date for gNya'-khri. This would put 619 years between these two kings. Allowing for a lifetime of 120 years, lHa-tho-tho-ri would have died in 494 A.D. This leaves somewhere between 100 and 120 years until the time of Srong-btsan-sgam-po in the seventh century.

THE FIRST APPEARANCE OF THE DHARMA

lHa-tho-tho-ri gnyan-btsan was born to King Khri-thog-rje-thog-btsan and Queen Ru-yong-bza' as one of four sons. We have not been able to discover anything about lHa-tho-tho-ri's early years, but one event late in his reign is well known. In the water-bird year of 433 A.D. (1552 years ago), lHa-tho-tho-ri at the age of sixty received the first sacred Buddhist texts to reach Tibet. Accounts say that the sky on that day was full of rainbows, and beautiful heavenly music could be heard. In a rain of flowers, a casket descended from the sky to the uppermost roof of the castle of Yum-bu-bla-sgang.

Opening the casket, the king found books written on golden leaves with lapis lazuli ink—the Za-ma-tog-bkod-pa'i-mdo (Karaṇḍavyūha-sūtra, Ny. 116) and the dPang-skong-phyag-brgya (Sakṣipūraṇasudraka, Ny. 267). The king also received the Cintāmaṇi, the sacred Six Syllable Avalokiteśvara Mantra, OṀ MAṆI PADME HŪṀ, carved on precious stones twelve inches high, and a golden stupa. That night in a dream, the king received a prediction that in five generations the one who could understand these things would be born.

lHa-tho-tho-ri himself did not comprehend the meaning of the texts or the sacred objects, but sensing their sacredness, he treated them reverently and called them the gNyan-po-gsang-ba, secret powerful objects. The king's faith and devotion to these secret objects was so strong that he was transformed: At the age of sixty, he became like a sixteen-year-old youth, and he lived for another sixty years.

When his son was old enough to reign, lHa-tho-tho-ri advised him to continue to honor the wonderful sacred objects, which would grant his every wish. Some say the king disappeared into the sky, though other sources say his tomb is in the Yar-lung valley at Dar-thang near the 'Phyongs-rgyas river. The bKa'-thang-sde-lnga gives a description of a treasury established by the royal family in honor of lHa-tho-tho-ri to benefit later kings.

The Chos-'byung-rin-po-che'i-gter-mdzod relates a tradition that two Buddhist translators appeared from Li-yul (Khotan) and Thod-dkar during the reign of lHa-tho-tho-ri. Once in Tibet, they invited an Indian abbot to come to teach the king. But as there was no system of writing, there was little possibility that the Dharma could spread among the people, and so they did not remain long. It seems possible that Central Asian Buddhists might have made their way into Tibet in lHa-tho-tho-ri's time. When the Chinese pilgrim Fa-hien traveled through Central Asia in 400 A.D., he found Buddhist centers flourishing at Lou-lan, Khotan, Kashgar, and Yarkand. Thod-dkar may refer to Tokharia west of Khotan, in the old Kuṣāṇa lands.

The Blue Annals mentions that the Paṇḍita Buddharakṣita (Blo-sems-'tsho) and the lo-tsā-ba Li-the-se stayed for a short time in Tibet during lHa-tho-tho-ri's reign and brought Buddhist texts with them. 'Gos Lo-tsā-ba finds this account more acceptable than the idea of texts falling down from the sky. This story, according to Nel-pa Paṇḍita, was just a way to explain the sudden appearance of these unusual objects to the Bon-pos close to the king, and 'Gos Lo-tsā-ba agrees with this view.

THE ERA OF LHA-THO-THO-RI

The era of lHa-tho-tho-ri seems to have coincided with the time of Kumārajīva, the great Central Asian Buddhist translator, who arrived in China in 401 A.D. He received patronage from the rulers of the Later Ch'in dynasty, founded by tribes modern scholars believe are related to the Ch'iang and Tibetans.

Some researchers suggest that the eastern parts of Tibet, as close neighbors and relatives to the Ch'in, were well acquainted with Buddhist teachings in the time of lHa-tho-tho-ri, and that northern and western Tibet had probably encountered the Dharma through monks or merchants traveling along the Silk Route. But it seems quite certain that there was no public royal patronage for Buddhism until several generations after the time of lHa-tho-tho-ri.

The first appearance of Buddhist influences in Tibetan culture corresponded in time with the introduction of the Dharma into Korea, and with the era of the great Indian master Vasubandhu, whom Tāranātha calls a contemporary of lHa-tho-tho-ri. In the previous century Christianity had been officially accepted in the Roman Empire, while the Muslim tradition was not to be established for another two centuries.

'BRONG-GNYAN-LDE'U'S SORROWFUL REIGN

lHa-tho-tho-ri's son was Khri-gnyan-gzung-btsan, but we know little of his reign. His queen was 'Brong-bza' Mo-dur-yang-gzher, and his minister was Shud-pu rGyal-to-re-nga-myi. When Khri-gnyan-gzung-btsan died, his tomb was built at Don-mkhar. There are interesting stories told about the son of this king. The son, 'Brong-gnyan-lde'u, took the throne and married a very beautiful woman, mChims-bza' Klu-rgyal from Dags-po. After being in Yar-lung for some time, she lost her beauty, for she was lacking special food eaten in Dags-po. With the king's encouragement, she made arrangements to obtain what she wished from her homeland. As the queen grew lovely once again, the king became very curious about this special food and opened the queen's storehouse. There he found a large supply of fish, which frightened and horrified him, for no one in Yar-lung ate fish in those days. The king fell quite ill from the shock.

One calamity followed upon another, for the son of 'Brong-gnyan-lde'u and mChims-bza' Klu-rgyal was born blind. When he grew older, his father gave him special counsel: to keep a pure heart, which is like gold; to listen to good advice, which is like nectar; and to renounce the samsaric mind, which is destructive. He urged him to respect and make offerings to his great grand-father's gNyan-po-gsang-ba, the secret objects lHa-tho-tho-ri had treated with such reverence.

Concerned that his blind son, now come of age, could not rule effectively as king, 'Brong-gnyan-lde'u instructed him to invite a Bon-po doctor from the neighboring land of 'A-zha. If the treatment failed, he was to contact the wife of a Bon-po from the land of Sum-pa, whose son should be brought into the royal family as an older brother and made the king.

'Brong-gnyan-lde'u then took all his personal possessions and those of the queen into the royal tombs at Yar-lung. There he and his wife would remain, as though dead, for the rest of their days. The government was to be left in the hands of the first minister until the treatment of the prince's eyes was attempted. Thus arranging everything, the king retired, knowing his son was well-protected, the instructions were clear, and his minister would be blameless in the eyes of the people.

STAG-RI BECOMES KING

The prince, nicknamed dMus-long, the Blind One, followed his father's instructions, worshipping the gNyan-po-gsang-ba and inviting the Bon-po doctor. After the treatment, the Bon-po returned home where his mother prophesied that in the future 'A-zha lands would come under the rule of Tibet. Soon after, the prince's sight returned during a trip to sKyid-shod where the first thing he saw was the form of a wild sheep on sTag-mo-ri mountain. Thus restored to good health, he became king, using the royal name sTag-ri-gnyan-gzigs, Seeing the Mountain Sheep. In sTag-ri's time, weights and measures for grain, butter, and other dairy products were developed and simple rules for business practices established.

sTag-ri married Gong-bza' sTong-btsun 'Gro-dkar and resided at 'Phying-ba. This was the center of the old kingdom, the land to which Gri-gum's son, sPu-de-gung-rgyal, had returned to rule. It is uncertain how far sTag-ri's influence extended, for in his day the land was beset with political intrigues. The Ngo-mtshar-rgya-mtsho notes that only about half of the twelve little kingdoms were under sTag-ri's control. How the kings had lost control over some of the original territories is not clearly explained in Tibetan histories or mentioned in the Tun-huang chronicles.

The Tun-huang chronicles do describe a power struggle during the rule of sTag-ri, for several prominent nobles resided in neighboring territories. Zing-po-rje sTag-skya-bo was lord of Nyen-kar, which might be the sTod-lung area northwest of lHa-sa. Zing-po-rje Khri-pang-sum was at Yu-sna in Ngas-po, a region perhaps near Kong-po, which seems to have included sKyid, probably the area around lHa-sa. The location of these old fiefs is not certain; they may have been in central Tibet, but it is also possible they extended into the northeastern borderlands.

INTRIGUES IN STAG-RI'S TIME

The lord of Nyen-kar, sTag-skya-bo, was a poor and unjust chief who never heeded advice and changed the laws to suit himself. Accounts say he took the good for bad and the bad for good, leaving the brave and the upright unrewarded, the simple folk frightened, and the ministers completely frustrated. Anyone who dared criticize him was severely punished. The people became confused, distrusting one another and their ruler, who finally lost the allegiance of his ministers.

The minister mNyan 'Dzi-zung warned his lord that if he continued to take the side of evil in all things, the government would no longer function, and the laws would lose their power. If the land is destroyed, the minister asked, who would be to blame? But sTag-skya-bo would not listen and stripped mNyan 'Dzi-zung of his rank. Burning with anger, minister mNyan joined forces with the lord of gNas-po, Khri-pang-sum, and killed his old lord. Khri-pang-sum took over sTag-skya-bo's fief, giving the lands of Yul-yel and Klum-ro to mNyan 'Dzi-zung.

But the people of Yul-yel and Klum-ro were not happy under the rule of the minister mNyan. His wife, a noblewoman from the Pa-tshab clan, was so unbearably haughty that finally a complaint was lodged with the gNas-po lord, Khri-pang-sum, by a member of the Myang clan. But Khri-pang-sum greatly admired the brave minister mNyan and would take no action. The people again were greatly discouraged.

There were other disputes which this lord did not settle to the liking of his people. For example, he once refused to allow repayment to the family of a murdered man because the king deemed the victim an evildoer. The lack of a clear code of laws seems to have caused great confusion and suffering among the Tibetan people, even when the rulers were considered good and honorable men.

The disgruntled clansmen of Myang and dBas joined up with one another and made their way to the castle of sTag-ri. Their secret plan is revealed in the Tun-huang chronicles in the form of a short poem, praising sTag-ri as a just ruler, one who knew how to handle the kingdom properly, just as a good rider knew how to handle his horse.

They sought out sTag-ri and swore with him a great oath of loyalty, and soon the clans of mNon and Tshe-pong joined them. Though sTag-ri's sister was married to Khri-pang-sum, he was willing to plan his brother-in-law's downfall and met with these clans secretly deep in the forest. But sTag-ri died before they could carry out their plan, leaving behind a wife and two sons, the older Slon-mtshan (Srong-btsan) and the younger Slon-kol.

A period of indecision and fear apparently followed. The subjects of sTag-ri had seen the finely dressed men going off to the forest with their ruler; they had watched them hurrying to the castle of 'Phying-ba-stag-rtse at night. But no one knew if these strangers were enemies or friends.

GNAM-RI TAKES CONTROL

An alliance ceremony was finally re-enacted between the rebellious clans and sTag-ri's son, Srong-btsan. The Tun-huang chronicles contain their oath, in which they disavow any allegiance to Khri-pang-sum and promised to join together with the king of sPu in the lineage of sPu-de-gung-rgyal. They pledged their loyalty, and even their lives, to the king alone, and swore to obey only his commands.

While his younger brother Slon-kol and their mother Queen sTong-btsun remained in 'Phying-ba, Srong-btsan, the new king, raised an army of ten thousand soldiers and conquered Khri-pang-sum. He forced his enemy's son to flee to the land of the Turks, and destroyed the castle at Yu-sna, changing the name of Ngas-po to 'Phan-yul. The ministers celebrated the victory with great festivities and song, calling the two brothers the divine kings, and giving Srong-btsan the new name of gNam-ri — Sky Mountain — for his empire was higher than the sky, his warrior's helmet more firm than a mountain.

gNam-ri distributed the rule of the subdued territories to the clans of Myang, mNon, dBas, and Tshe-spong. He granted them the rule over families to work their lands, and they became official ministers within his government. A political network thus spread power among the clans that supported the king. He took a noblewoman of the Tshe-spong clan named 'Bri-bza' Thod-dkar as his wife.

During these years, several ministers were especially famous, though it is not certain which kings they served. The Tun-huang documents name a member of the mGar clan who had knowledge so deep, it was said that he could read the minds of three people at once. Another was also psychic, capable of replying to the query of an envoy before the envoy had spoken his message.

Several skilled and fearless ministers were known to have aided gNam-ri. Khyung-po sPung-sad-zu-tse was an aggressive supporter of the king. Zu-tse had seized the lands of rTsang-bod and offered them, together with their twenty thousand families, to gNam-ri. Zu-tse also disposed of a chief known as Mong sNgon-po, who had grown hostile to the king and his brother. When Dags-po revolted, another minister, Son-'go-myi-chen, crushed the prince of Dags-po for gNam-ri. Though he was an inexperienced commander, he showed tremendous bravery, and the king gave him lands and cattle as a reward.

But the intrigues that had put gNam-ri in power did not come to an end during his reign. The minister Khyung-po Zu-tse grew jealous of the rising power of other ministers and generals. At a great festival held to celebrate the selection of a new minister, Zu-tse complained bitterly to the king. The Tun-huang chronicles preserve his long song in which he reminds the king that he, Zu-tse, had been the one to rid the kingdom of the "tiger of Mong" and the "vulture in rTsang-bod."

The king heard him out, hoping one of the ministers would reply. Finally, the new minister Zhang sNang of the Myang clan composed a few verses, reminding the proud Zu-tse of the efforts made by others. The real credit, Zhang sNang insisted, should go to the penetrating leadership of the king.

"Without the arrowhead of iron,
the bamboo shaft will pierce nothing.
Without the sharp and pointed needle,
the finest thread will sew nothing."

The extent of gNam-ri's power is not certain. The Tun-huang documents make it clear that Sum-pa, Dags-po, Kong-po, Myang-po, and some Zhang-zhung tribes were subjects of gNam-ri, and that the tribes and clans of lDong, sTong, bSe, Khyung, lHo, and rNgegs were his subjects as well.

FOREIGN RELATIONS AND TRADE

The bKa'-thang-sde-lnga reports that the reign of gNam-ri was a time of continuous strife. It notes that the Turks and the Chinese were both resisted, but Tibet's specific dealings with them at this time are not documented. By 552 two powerful Turkish khanates had been established by the sons of Bumin Khagan: the Eastern Khanate in Mongolia and the Western Khanate in Dzungaria, at the northwestern end of Central Asia. The Eastern Turks made alliances with China, while the Western Khan Istami (r. 552–575) made alliances first with the Persian ruler Chosroes I and with the Byzantine Emperor Justinian II. What contact gNam-ri might have had with either khanate is uncertain, except that Tun-huang records do note he drove his enemy's son to flee to the Turkish lands.

North China in this period was under the rule of the Turkish-Mongol Toba Wei dynasty (386–534) until the Sui dynasty re-united north and south China in 589. No details about struggles with China are given by Tibetan historians. But there was some cultural contact between the two countries, for Tibetan histories assign the first use of Chinese medicine, astrology, and calculation techniques to the time of gNam-ri.

T'ang dynasty records indicate that the first official contact between Tibet and China took place under gNam-ri's son. According to Han dynasty annals, however, trade had been going on between China and the tribes in the northeast since the time of the Han dynasty (220 B.C.–202 A.D.). Large trade marts had been in operation for centuries in the Koko Nor region and to the northeast. Horses from the Tibetan plateau were highly valued; gNam-ri himself is said to have possessed a horse famous for its unusual intelligence.

Salt was also an important item of trade, and quantities of it were collected from the salt flats on the northern plain of Byang-thang. A story is told of a particular hunting expedition gNam-ri made in the lands around Dang-ra lake. He had placed the carcass of a wild yak across his horse during the journey across the salt flats back to Yar-lung. Upon his return, the meat had become salty and very flavorful. From that time on, salt was widely used.

gNam-ri's reign came to an end much the way it had begun when a conspiracy of discontented nobles managed to poison him. He left behind two sons and a daughter, the oldest boy an unusual child named Srong-btsan. Though gNam-ri had won the allegiance of the most important Tibetan clans and controlled a large territory, the real work of founding the Tibetan Empire was to be accomplished by his son Srong-btsan-sgam-po. gNam-ri's son would also bring the Dharma into Tibet, fulfilling the prediction given to lHa-tho-tho-ri five generations before.

lHa-tho-tho-ri gnyan-btsan

THE ROYAL LINEAGE OF THE TIBETAN KINGS AND QUEENS

THE SEVEN HEAVENLY KHRI

gNya'-khri-btsan-po and gNam-mug-mug
Mu-khri-btsan-po amd Sa-ding-ding
Ding-khri-btsan-po and So-tham-tham
So-khri-btsan-po and Dog-mer-mer
Mer-khri-bstan-po and gDags-kyi-lha-mo-dkar-mo
gDags-khri-btsan-po and Srib-kyi-lha-mo
Srib-khri-btsan-po and Sa-btsun Rlung-rje

THE TWO STENGS OF THE UPPER REGION

Gri-gum-btsan-po and Klu-srin-mer-lcam
sPu-de-gung-rgyal and 'O-ma-thang sMan-mtsho

THE SIX LEGS OF THE EARTH

E-sho-legs and rMu-lcam Bra-ma-na
De-sho-legs and Klu-sman Mer-mo
Thi-sho-legs and bTsan-mo Gur-sman
Gong-ru-legs and mTsho-sman 'Brong-ma
'Brong-gzher-legs and gNam-sman Bu-mo
I-sho-legs and rMu-lcam dMar-legs

THE EIGHT LDE KINGS

Za-nam-zin-lde and mTsho-sman Khri-dkar
lDe-'phrul-nam-gzhung-btsan and Se-gnyan Mang-mo
Se-snol-gnam-lde and Klu-mo Mer-mo
Se-rnol-po-lde and'O-za sDe-mo-mtsho
lDe-snol-nam and Khri-sman rJe-mo
lDe-snol-po and Se-btsun gNyan-mo
lDe-rgyal-po and sMan-btsun Lug-gong
lDe-spring-btsan Nye-btsun Mang-ma-rje

THE FIVE BTSAN KINGS

To-re-long-btsan and 'O-ma lDe-bza' Khri-btsun-byang-ma
Khri-btsun-nam and sMan-bza' Khri-dkar
Khri-sgra-spungs-btsan and sMan-gsum Klu-steng
Khri-thog-rje-thog-btsan and Ru-yong-bza' sTong-rgyal-mo-tsho
lHa-tho-tho-ri gnyan-btsan and rNo-bza' Mang-dgar

CRISIS IN EURASIA

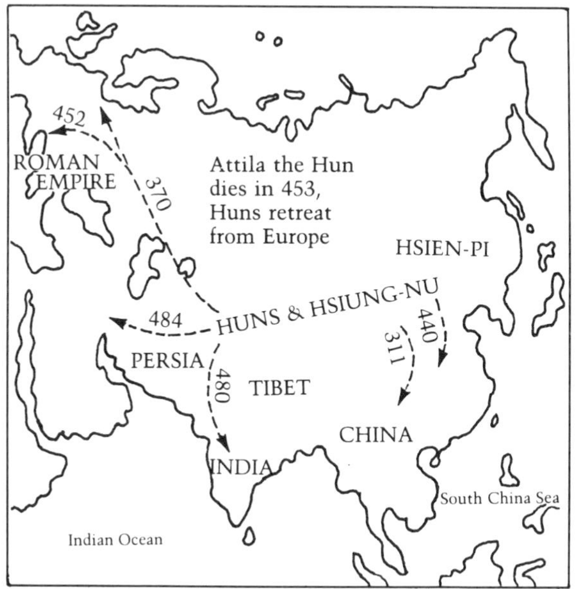

The fourth and fifth centuries in Eurasia witnessed the collapse of the ancient world order as nomad tribes from the steppes of Central Asia swept into India, Persia, China, and Europe. The Gupta Empire and the Roman Empire fell, and China was dominated for nearly three centuries by foreign dynasties established by Turkish and Mongol tribes, and by tribes said to be related to the Tibetans, such as the founders of the Ch'in and the Later Liang dynasties.

NEW EMPIRES TO THE NORTH

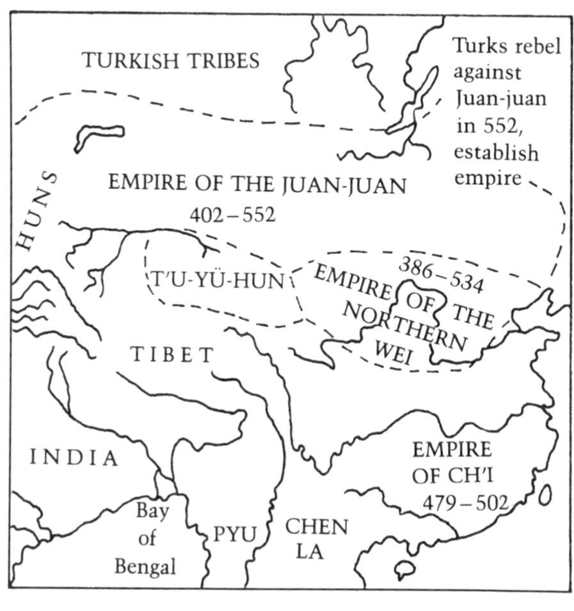

By the sixth century, several Turko-Mongol empires were established north of Tibet by the Juan-juan, the To-pa Wei, and the T'u-yü-hun. The T'u-yü-hun became known in Tibetan records as 'A-zha, while the Juan-juan, or their successors, the Turks, may have been the tribes with whom the son of King gNam-ri's defeated rival took refuge. gNam-ri is said to have had dealings with the Turks and with China, but little is known about Tibet's foreign affairs in this era.

THE ERA OF LHA-THO-THO-RI

TIBET	INDIA	CENTRAL ASIA	WEST
374 A.D. Birth of lHa-tho-tho-ri, last of Five bTsan	c. 300 A.D. Indianized states in Southeast Asia flourish	c. 300 A.D. Oasis states flourish; Buddhist centers well-established	313 A.D. Edict of Milan: Christianity tolerated in Roman Empire
433 A.D. lHa-tho-tho-ri receives Buddhist texts, sacred objects, mantra; king transformed into young man, lives 60 more years	c. 300–550 A.D. Vākāṭakas rule south India	301–589 A.D. China divided into many dynasties; some founded by Hsiung-nu, Hsien-pi, and Ch'iang/Tibetan tribes	330 A.D. Capital of Roman Empire moved to Constantinople
c. 495 A.D. Khri-gnyan-gzung-btsan rules at Yar-lung	320 A.D. Beginning of Gupta Era: peace and prosperity across north India; end of republican states	301–439 A.D. Sixteen Kingdoms period in China	354 A.D. Birth of St. Augustine
c. 520 A.D. 'Brong-gnyan-lde'u rules at Yar-lung; marries mChims-bza'	376–415 A.D. Candragupta II: height of Gupta Empire; Śakas expelled from western India	313 A.D. T'u-yü-hun tribes settle north of Koko Nor; control up to Khotan by 445	360 A.D. Huns enter Europe
c. 540 A.D. sTag-ri-gnyan-gzigs rules at Yar-lung; allies with faction of neighboring nobles: dBas, mNon, Tshe-spong, Myang	460 A.D. King Skandagupta defeats Huns in western India; Huns control Gandhāra	317–589 A.D. Period of Northern and Southern Dynasties divides China	410 A.D. Visigoths sack Rome
c. 560 A.D. gNam-ri and his brother renew alliance	c. 500 A.D. Toramāṇa, barbarian chief, presumably Hun, controls northwest India, Kashmir, Punjab	350 A.D. Huns invade western Central Asia	455 A.D. Vandals sack Rome
gNam-ri conquers Ngas-po, Klum-ro, Nyen-kar, and Dags-po	c. 500–600 A.D. Advances in sciences: logic, astronomy, algebra, arithmetic, epistemology	344–413 A.D. Buddhist master Kumārajīva; translator in China	470 A.D. Huns withdraw from Europe
gNam-ri marries 'Bri-bza' Thod-dkar, has two sons; gNam-ri poisoned by conspirators	c. 530 A.D. Bālāditya defeats Mihirakula, son of Toramāṇa	433 A.D. Attila rules Huns	476 A.D. End of Western Roman Empire
		484 A.D. Hun empire expands	486 A.D. Frankish kingdoms founded by Clovis
		552 A.D. Turkish Empire established: controls Silk Route	507 A.D. Franks defeat Visigoths
			527 A.D. Justinian I rules Byzantium, Eastern Roman Empire
			530 A.D. Possible date for King Arthur

PART THREE

THE EMPIRE

PERSPECTIVE

AN OVERVIEW OF HISTORY

The era of the Dharma Kings is an especially important time in Tibetan history, for we witness long-lasting accomplishments in the making. From a culture that was undoubtedly ancient and complex, but certainly not cosmopolitan, urban, or literate, a sophisticated and powerful state emerged in the space of a few generations.

Though Srong-btsan-sgam-po's father was hailed as a wise and just leader, his influence was very limited. His son commissioned the creation of an alphabet, laid the foundations for the Buddhist tradition, and introduced Tibet to the arena of world powers, where its political and military decisions affected regions far beyond its frontiers. Succeeding kings continued to build on these accomplishments. The Tibetan court began sponsoring art and architecture on a grand scale. Huge translation projects produced work of such high quality that modern translators could not begin to equal it.

The Buddhist tradition calls these kings Bodhisattvas, incarnations of Avalokiteśvara, Mañjuśrī, and Vajrapāṇi. However we understand them, it is clear they ignited the genius of the Tibetan people. During their reigns, outstanding individuals arose in all walks of life — ministers, queens, generals, scholars, craftsmen, monks, mystics, artists. In retrospect, it is clear that they had tremendous impact on Tibetan culture; actions they undertook over a thousand years ago continue to have an effect in our own times.

Though the empire of the Dharma Kings finally dissolved, the seeds of the Buddhist teachings had been successfully planted among the Tibetan people, who from that time on never lost their desire to follow the Dharma. The influence of Buddhism upon Tibetan civilization is hard to measure, for the teachings permeated the old culture. The Tibetan Buddhist civilization born from this marriage was multifaceted and unique—a blend of the most advanced doctrines in the world with the native Tibetan spirit. The civilization that blossomed in the era of the Dharma Kings has endured for over thirteen hundred years, while the older heritage it encompassed is commonly traced back nine hundred years to the time of the first king in the third century B.C. If we consider also the history of the ancient tribes, it appears that the very beginnings of Tibetan culture extend back beyond the time of the Buddha into prehistory.

During several thousand years of development, Tibetan culture has remained intact, though leadership has shifted from time to time, and political fortunes have changed. The dynasty of kings beginning with gNya'-khri-btsan-po ruled a kingdom whose boundaries are uncertain; by the time of the thirty-third king, Srong-btsan-sgam-po, the kingdom was becoming an empire. By the time of the forty-second king, the empire had collapsed, and political power was spread among a number of influential leaders around the country.

If we were to quickly summarize later events beyond the scope of this book, we would see such shifts continuing to occur. In the thirteenth century Tibet, as well as China, Korea, Central Asia, Persia, and Russia, became part of Genghis Khan's Mongol Empire for about a century. In this period, the Sa-skya-pa hierarchs became the dominant political power in Tibet. In the fourteenth century, the Phag-mo-gru dynasty severed ties with the Mongol Empire. In the fifteenth century, political power was held by the Rin-spungs-pa princes, and in the sixteenth century by the gTsang-pa princes. The government of the Dalai Lamas was established in the middle of the seventeenth century. In 1959 the Dalai Lama left central Tibet, and the Communist regime of China took over the government.

Such political changes inevitably have an effect upon the development of a culture. Certain choices are made—new elements are added to the culture while some older ones are left behind. These decisions may be made by the people as a whole, or perhaps through the influence of a few, depending on the distribution of power and the temper of the times. We have seen that changes occurred under the leadership of gNya'-khri-btsan-po and Gri-gum-btsan-po. The next chapters describe developments in the era of the Dharma Kings. Succeeding generations continued to modify the culture, but its essential elements remained deeply rooted in Tibetan language and customs, and in the spirit of the Tibetans themselves.

A general overview of history reveals that the continuity and integrity of an ancient civilization can persist despite changing circumstances. But history also shows that a heritage can be lost, if people relinquish knowledge of the past.

With each changing circumstance, people are faced with the need to determine how to preserve the values and knowledge of the past, while introducing new practices and ideas. Sometimes old traditions are so strong they cannot easily be altered, either for better or for worse. In previous centuries, the slower rhythm of change supported the continuity of older ways in the face of new circumstances, and also provided more time for new ideas to grow naturally.

But in today's world, rapid change affects all aspects of life—economic, political, social, technological. Its momentum creates a pressure upon modern people to keep up with the current. Faced with so much new to master, people tend to lose touch with the past. Exciting new perspectives appear to be more advanced or more scientific; older ways of life seem to be "backward" or "mistaken." Confronted with the variety of lifestyles and philosophies available in the modern world, younger generations who no longer identify with their own heritage have little basis for their choices. When a whole culture renounces its heritage, actions considered unacceptable in the past may today be considered heroic, but tomorrow, lamentable.

There seems to be no guarantee that modern cultures are proceeding steadily toward the perfect society. As older ways of life are forsaken, we may rid ourselves of one set of defects, only to find ourselves faced with a more complex set of problems. Drawbacks and advantages are interwoven into both ancient and modern civilizations. Given enough time, we may be able to sort through the best of our human heritage and thus gradually perfect human knowledge.

If we abandon our knowledge of the past, then the human heritage will gradually slip from memory all over the world—and with it a precious reservoir of wisdom that belongs to all people. The fuller our knowledge of the past, the more likely we are to guide the future with intelligence, without repeating old mistakes.

History is full of the evidence of human mistakes. Looking at what is already past, we can often see the errors that caused great suffering or allowed something valuable to be lost. But to see which present actions are best is much more difficult, especially in modern times when changes occur before the situation can be carefully evaluated. Once rapid and dramatic change has begun, conflicting camps arise with different opinions and doctrines. Holding to a doctrine makes it difficult to understand alternate perspectives and develop a wide, clear view. Competition, that age-old manifestation of human selfishness, comes into play. Together, ignorance and power-seeking begin to distort human values and our vision of truth. When objectivity does not seem possible, or when information is readily distorted, there is little possibility of predicting the future. It may be necessary to wait for the situation to bear fruit before passing judgment.

Even when a clear judgment is possible, there may be no benefit from pronouncing it, or no way to implement it. The wars and instability that plague modern times may remind us that traditional values and the slower pace of the past have their advantages. But the forces moving modern civilization are strong and pervasive.

In the last half century older cultures previously untouched by dramatic change have become part of a worldwide process of upheaval — no corner of the globe remains isolated. The encounter between modern culture and older civilizations seems more complex than the rivalry between neighboring cultures in the past. Ancient patterns of interaction seem to be dissolving while new ones are not yet clear. North and South, East and West have not only met, but have begun to affect each other in ways that are difficult to perceive from our limited human viewpoint. In this new complex and unpredictable arena, the old selfish patterns are being played out with increasingly powerful technology. It is no wonder that serious questions are being raised about the direction of modern civilization.

It seems too soon to say whether the choices being made by modern people all over the world will prove beneficial in the long run. But if human intelligence is not overridden by selfishness or confined by doctrine, then perhaps we will find a way to resolve disparate points of view. If our energy can be rescued from conflict and anxiety, it may be turned toward the realization of the best destiny of the self and the best benefit for humanity. Instead of being a source of conflict, the wealth and power of the modern world could be used for the common good. Human intelligence operating freely could support us through this period of transition while the new is being born from the old.

In these times, it seems especially important to know the past, not only to study its lessons, but also to preserve its knowledge. Faced with the complex difficulties of modern times, we cannot afford to discard such a rich heritage. If the wisdom and experience of older civilizations were joined to the store of modern expertise, the fully mature body of humanity's knowledge could become the foundation for world harmony.

EMERGENCE OF THE MODERN WORLD

1453
Ottoman Turks seize Constantinople: Byzantine Empire ends

1492
Columbus discovers America

1532
Spain conquers Incas

1571
Portuguese colonies in Africa

1581
Russia begins conquest of Siberia

1607
English settle in America

1610
Scientific advances begin in Europe

1619
Dutch colonies in Indonesia

1642
Fifth Dalai Lama takes power in Tibet

1650–60
Dutch and French found colonies in Africa

1770
Rapid technological advances in Europe

1789
French Revolution

1818
British control India

1842
Britain annexes Hong Kong

1846
Western exploration of Tibet begins

1848
Communist Manifesto by Marx and Engels

1850
T'ai-p'ing Rebellion in China

1859
Theory of evolution published

1863
French control Indochina

1864
American Civil War

1875
Growth of socialist parties in Europe

1895
Thirteenth Dalai Lama assumes power

1900
Freud's theory on dreams published

1900
Boxer Uprising in China

1905
Einstein's theory of relativity

1910
Chinese troops enter lHa-sa

1910
Mexican Revolution

1911
Chinese Revolution Sun Yat-sen is President

1914
World War I

1917
Russian Revolution

1921
Reza Khan takes control of Persia

1922
Last Ottoman ruler deposed in Turkey

1930s
Revolutions and civil wars begin in Latin America and South America

1933
Hitler rises to power

1934
Long March of Chinese Communists

1939–1945
World War II

1945
Cold War begins

1947
India and Pakistan become independent

1949
Communists win control of China

1950
Communist Chinese troops enter Tibet

1950
Korean War

1954
Laos, Vietnam, and Cambodia independent

1954
African independence movements begin

1955
Warsaw Pact: Communist control of eastern Europe

1959
Cuban Revolution

1959
Dalai Lama leaves Tibet

1966
Cultural Revolution in China

1975
Communists control Vietnam, Laos, and Cambodia

CHAPTER TWELVE

ADVANCES IN TIBETAN CIVILIZATION

The foundation of the Tibetan Empire and a firm basis for the Tibetan Buddhist tradition were both established in the the seventh century by Srong-btsan-sgam-po. During his reign, the Tibetan culture began to develop into a cosmopolitan civilization equally at home with the sophisticated thought of the Buddhist tradition and the complexities of medieval Asiatic politics.

Srong-btsan-sgam-po was the son of King gNam-ri-srong-btsan and Queen Tshe-spong-bza' 'Bri-bza' Thod-dkar. He was born at rGya-ma in Mal-dro, a region northeast of present-day lHa-sa, in the castle known as Mi-'gyur Byams-pa-gling. Histories generally agree that his birth took place in an ox year, but which one is disputed (see chapter 14). The Tibetan tradition speaks of him as an incarnation of the Bodhisattva Avalokiteśvara, purposely manifesting in human form to lay the foundation for the Dharma in Tibet. Srong-btsan-sgam-po's reign was predicted in the Mañjuśrīmūlatantra: As the Dharma spread to the north, a king would arise in lHa-ldan, a descendant of the Licchavis and a great siddha.

Buddhist histories describe how the skies were full of unusual beauty on the day of Srong-btsan-sgam-po's birth. All the Buddhas and Bodhisattvas came to bless this unusual child, who was specially marked with a flame-shaped tuft on his head, the sign of a great man. At the age of thirteen, Srong-btsan-sgam-po was enthroned in great ceremony. Buddhas and Bodhisattvas again gave him their blessings. Samantabhadra initiated him, and the actual form of Amitābha arose on the top of the king's head. This enthronement and initiation scene was painted in gold upon the walls of the Khra-'brug temple later founded by Srong-btsan-sgam-po in Yar-lung.

A WRITING SYSTEM FOR TIBET

Soon after his reign began, Srong-btsan-sgam-po saw the need for a government guided by clearly established laws to stabilize the country and promote moral behavior. A system of writing was essential for this purpose, as well as for bringing the teachings of the Dharma to the people of Tibet. Srong-btsan-sgam-po gave his trusted minister Thon-mi Saṁbhoṭa the task of devising such a system, sending him abroad to work with foreign scholars.

Thon-mi Saṁbhoṭa sought out an expert in the Indian literary language, a scholar named Li-byin, according to the rGyal-rabs-gsal-ba'i-me-long; some sources call him Legs-byin. Offering gifts of gold to the learned master, Thon-mi studied two different Sanskrit grammars, as well as poetry and logic.

As Thon-mi became learned in many different scripts, he created an alphabet to represent the sounds of spoken Tibetan. Choosing from among the fifty Indian letters, he adapted thirty symbols for consonants and devised variant forms for use in combinations. Three "head letters" were to be written as superscripts; four "tail letters" were to be written as subscripts. From among the fourteen vowels of Sanskrit, he selected four signs to be written above or below the consonants, while each consonant was to contain an inherent "a," just as in Sanskrit. Some letters were modified to represent sounds not existing in the Indian languages, while signs for sounds not used in Tibetan were omitted. Thon-mi worked out spelling rules for combining letters and determined which consonants could serve as prefixes and suffixes.

The rGyal-rabs also notes that Thon-mi worked with the Paṇḍita lHa-rigs-seng-ge (Devavidyāsiṁha), becoming expert in all the treatises on grammar, as well as in the five traditional sciences: language, logic, philosophy, medicine, and art. The name Devavidyāsiṁha may be connected with one of three Kashmīri teachers of this era who are mentioned by the Buddhist historian Tāranātha: Vidyāsiṁha, Devavidyāsiṁha, and also Devasiṁha. This seems to indicate that Thon-mi went to Kashmir, and Bu-ston and the La-dwags-rgyal-rabs agree. But the Blue Annals reports that he studied in India, which was a separate kingdom in those days. The rGyal-rabs-gsal-ba'i-me-long states that he met Li-byin in India, though it does not mention where he worked with lHa-rigs-seng-ge.

The Indian Nāgarī script is often mentioned as the model for Tibetan; some sources also state that the Lantsa script was the inspiration for the Tibetan dbu-can printed script, and the Vartula script was the model for the Tibetan dbu-med cursive script. Modern scholars, however, believe that these particular Indian scripts came into use after the seventh century, and they suggest instead the northwestern variety of the Gupta script as a likely model. After comparing inscriptions in Gupta script with samples of ancient Tibetan script, the author of the Deb-ther-dkar-po concluded that the Gupta script was the model used.

Thon-mi Saṁbhoṭa returned to Tibet not only with a new alphabet, but also with Mahāyāna Sūtras and Tantras. In a great celebration he presented the script to Srong-btsan-sgam-po in the form of an unusual piece of poetry in honor of the king. The first four lines were composed of words using only the vowel "a"; the fifth line contained only words using the vowel "e"; the sixth line used only words with "i"; the seventh used only "o" and the eighth only "u". This lovely poem praised the spiritual qualities of the king whom Thon-mi deeply admired.

A sample of the writing shown by Thon-mi Saṁbhoṭa to Srong-btsan-sgam-po was erected just north of lHa-sa behind the nine-storied castle of Pha-bong-kha, which was built by the king as his own meditative retreat. There upon a large rock Oṁ Maṇi Padme Hūṁ was engraved in the new Tibetan script.

The great attention being paid to Thon-mi Saṁbhoṭa began to cause uneasiness among the other ministers. So Thon-mi gathered the ministers together and explained how he had undergone great physical hardship in his travels, how extensive his studies had been, and how all his energy had been devoted to devising the alphabet. Indeed, Thon-mi's genius had been recognized by the Indian masters he had worked with, and as the ministers listened to his adventures, they began to understand what an unusual man he was. When Thon-mi declared, "I am the first learned man to arrive in this uncivilized country! There is no one like me," the others knew that his pride was truly selfless—his efforts were for the benefit of his people.

Working in the temple of Me-ru in lHa-sa, Thon-mi composed eight treatises on Tibetan grammar and writing that laid out the newly devised rules, together with examples and explanations. Only two of these treatises have survived to the present day, the rTags-kyi-'jug-pa and the rTsa-ba-sum-cu-pa.

THE BENEFITS OF THE NEW SCRIPT

It seems possible that earlier efforts to create a written language could also have been made, and that some of the Tibetan tribes had developed writing systems for their own dialects. There is mention in the T'ang annals, for example, that some of the Ch'iang tribes used Indian scripts, and ancient Zhang-zhung is said to have had a written script. But it is clear that no national language existed before the days of Srong-btsan-sgam-po. With the establishment of standard grammar and spelling, all literate persons could communicate with one another and with future generations. Unlike Europe, where records, histories, philosophy, and literature were all composed in classical Latin, Tibet transformed its spoken language into a "classical" form. The spoken languages of most European peoples were not regularly used even for record-keeping until the fourteenth century.

The Tibetan language, already very old by the time Thon-mi developed these standardizations, contained words whose ancient meanings gradually became uncertain or were forgotten. Yet many very old terms are still in use, for the basic sounds and rhythms of Tibetan have carried on into modern times. Today, Tibetan-speaking people can read material from the era of the Dharma Kings, especially texts revised or composed in the ninth century. English-speaking people, on the other hand, can scarcely decipher texts in tenth-century Old English, which appears to the untrained eye to be a completely different language.

THE FIRST TIBETAN LITERARY WORKS AND TRANSLATIONS

In order to master the new script and grammar and to begin translations, Srong-btsan-sgam-po retired for four years. Thon-mi Saṁbhoṭa and Srong-btsan-sgam-po together translated twenty-one Avalokiteśvara texts, as well as the Za-ma-tog-bkod-pa (Karaṇḍavyūha, Ny. 116) and the dPang-skong-phyag-brgya-pa (Sākṣipūrṇamsudraka, Ny. 267), which King lHa-tho-tho-ri had received long before; the 'Phags-pa-dkon-mchog-sprin (Ratnameghasūtra, Ny. 231) and the 'Dus-pa-rin-po-che-rtogs-gzungs were also translated at this time.

Other translators working with the king and Thon-mi included Brāhmaṇa Śaṁkara, the Nepalese Śīlamañju, the Indian teacher Kusara, the Chinese Hwa-shang Mahādeva, Ācārya Dharmakośa, who was a student of Thon-mi Saṁbhoṭa, and lHa-lung dPal-gyi-rdo-rje. Word of Srong-btsan-sgam-po's efforts for the Dharma reached Khotan, and two Khotanese monks traveled to lHa-sa to study several years with the king.

A great Dharma teacher and scholar in his own right, King Srong-btsan-sgam-po composed the Maṇi bka'-'bum, a compendium of various chronicles, Avalokiteśvara sādhanas, and detailed instructions for meditation practices. The king taught privately, instructing certain of his subjects in special practices, especially those related to Avalokiteśvara. Thon-mi Saṁbhoṭa, whom the Buddhist tradition considers an incarnation of Mañjuśrī, also taught eighty great disciples, who eventually became successful yogins.

The king's texts were eventually collected together and copied in gold ink on blue silk cloth. These were carefully hidden in three locations where they were found in later centuries by sGrub-thob dNgos-grub, mNga'-bdag Nyang, and Śākya 'Od. Old records note that Srong-btsan-sgam-po also preserved many treasures and precious objects by concealing them in precisely chosen locations.

Although Srong-btsan-sgam-po was deeply devoted to the Buddhist teachings, he treated the Bon-pos fairly. According to the Bon text Srid-pa'i-rgyud-kyi-kha-byang-chen-mo, he allowed Bon teachings to be written down in special turquoise ink. In his own writings, Srong-btsan-sgam-po notes that he himself buried a number of these texts to preserve them for the future.

Through Srong-btsan-sgam-po's and Thon-mi Saṁbhoṭa's careful efforts, the Tibetan language was being deliberately crafted into a tool to convey sophisticated concepts with precision. Within a century and a half after the foundation laid by Srong-btsan-sgam-po, large teams of Tibetan lo-tsā-bas were working with Indian paṇḍitas to translate hundreds of Buddhist texts dealing with subtle mystical subjects, complex meditation techniques, and refined philosophical concepts.

CREATION OF THE CONSTITUTION

The new script also allowed records, genealogies, legends, and poetry to be preserved in writing, and made possible a written code of laws. One of the most valuable gifts Srong-btsan-sgam-po gave the Tibetan people was the Six Codices of the Tibetan constitution, which set forth all the laws and principles followed by the Tibetan government for centuries afterward.

The Tun-huang documents note that in the time of this king, the ranks of officials and their respective powers were established; fields and pastures were divided, water was apportioned, and measures and weights were standardized. Later histories contain extensive copies of old records on the constitution. Some of this material is difficult to understand, involving governmental offices and ranks that are not completely clear and place-names that are unfamiliar. Study of the numerous Tun-huang documents referring to military units and administrative matters may help clarify some of these details; nevertheless, the outlines of the constitution are clear.

The king first established additional ministries to serve the Tibetan state. Over three hundred ministers were eventually selected, including four great ministers, six ministers of the interior, six of the exterior, and a special council. While the king retired to work on treatises and the constitution, all state affairs were handled by the senior ministers. Unfamiliar with such complete delegation of royal power, people began to grow uneasy and wondered who was in charge—the ministers or the king. Srong-btsan-sgam-po soon gave a public address in which he explained that the ministers' special powers had been granted them by the king for his own good purpose, the writing of the constitution.

THE SIX CODICES OF THE CONSTITUTION

The first of the six codices laid out the administrative and territorial divisions in Tibet: the internal divisions known as ru; the political divisions in the mDo-khams region; the military divisions and the civilian districts; the ministers who held the council; and the special army divisions guarding the frontier.

The Tibetan army was a huge institution including sixty-one divisions of a thousand men each. Each ru had eight of these sixty-one divisions plus two additional divisions, a smaller unit, and an Imperial Guard unit, bringing the total to ten in each of the four directions. Zhang-zhung and Sum-pa contributed ten and eleven units respectively, making sixty-one in all.

The frontiers were guarded by six divisions placed in three strategic points. The Turks were held back by troops stationed in the far northwest. The Middle Division guarded the border with

'Jang, the Nan-chao kingdom to the southeast of Tibet. The Lower Division guarded the borders with China in the northeast.

Six additional bureaus were established: a military office responsible for defense; an office of the cavalry in charge of arrangements for transporting troops; an office of the treasury, which kept track of grain, gold, and silver supplies; an agricultural department especially concerned with animal husbandry; and a justice department that made legal rulings following the code of laws included in the constitution.

All of these arrangements together made up the first part of the first codex of the constitution. The first codex also included the "thirty-six institutions," a collection of guidelines, ranks seals, and insignia that detailed the workings of the government.

The six insignia established for Tibetan officials included special badges in gold, turquoise, silver, copper, and iron, as well as metal alloys like bronze. Each type was created in both a large and small size, making twelve insignia in all. Ministers, religious teachers, king's attendants and guards, leaders of the clans, military commanders, and outstanding military heroes each received a specific rank and insignia. The T'ang annals show that the T'ang Empire was especially aware of the ranking system and insignia used by the Tibetan officials who dealt in foreign policy.

The second great codex of the constitution laid out the standard weights and measures for grain, gold, and other items in order to regularize business practices and trade. The third codex included the moral laws of the land, the fifteen regulations of the kingdom, the sixteen pure guidelines based on the Dharma, and the ten nonvirtuous actions that should be prohibited or discouraged. The last three codices of the constitution contained examples and discussions of how the laws were to be applied.

The Tibetan constitution gave the people an appreciation for written codes of law and a clear vision of how society could operate. It is interesting that the era of Srong-btsan-sgam-po was also the time of the Japanese Prince Shōtoku, who established a constitution based upon Buddhist and Confucian principles. In

Europe, Roman law already had been codified centuries earlier, but the laws of the Germanic and Frankish tribes were just being put into Latin in the fifth, sixth, and seventh centuries. It was not until the ninth and tenth centuries that the foundation was laid for the modern European nations, such as Germany, Italy, and France. These countries did not have written constitutions until the eighteenth and nineteenth centuries. England's Magna Carta, which gave the nobles the formal right to advise and confer with the king, was not written until the thirteenth century.

Because of its ministerial councils and codes protecting the rights of the individual, the constitution written by Srong-btsan-sgam-po is compared by some scholars with the ancient republican government of the Licchavis, to whom the Tibetan kings are often said to be related (see chapter 9). In any case, the Tibetan constitution is a remarkable document that stood the test of time. The principles established by this king were later to serve as the model for the Tibetan government from the fourteenth century until very recent times.

Chos-rgyal Srong-btsan-sgam-po

ANCIENT SCRIPTS

Writing in the form of hieroglyphics, cuneiform, and pictographs was first developed in several centers of civilization in the ancient world beginning about 3000 B.C. Eventually alphabetic scripts arose in the Middle East and spread to the Greeks and Romans. Some of the Indian scripts may have been modelled on an alphabet from the Middle East; some scholars think Indian Brāhmī script developed from the ancient Harappan script, which has not yet been deciphered.

INDIAN SCRIPT FOR TIBETAN

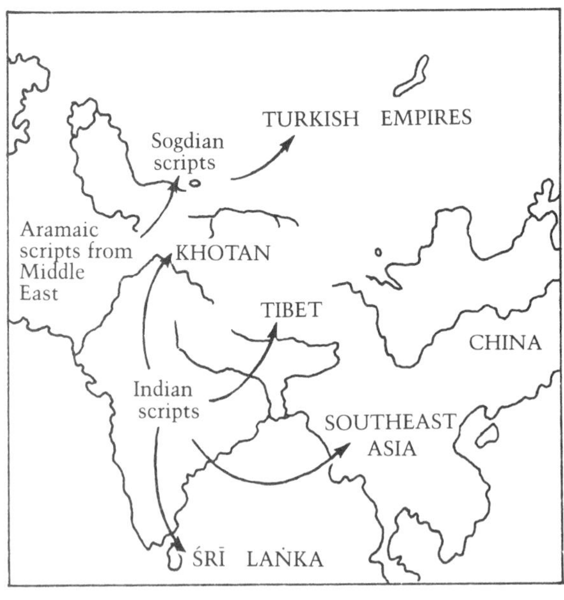

The Tibetan alphabet was based upon one of the Indian scripts in use during the seventh century A.D. when Srong-btsan-sgam-po commissioned chief minister Thon-mi Saṁbhoṭa to devise writing for the spoken language of Tibet. Ancient Indian scripts included Kharoṣṭhī, Brāhmī, and related scripts such as those in the Gupta style. Indian models were also used to develop scripts for languages in Southeast Asia, Śrī Laṅka, and in some Central Asian states.

THE DEVELOPMENT OF WRITING AND LAW

DEVELOPMENT OF WRITING	DEVELOPMENT OF WRITING	LAWGIVERS CONSTITUTIONS	LAWGIVERS CONSTITUTIONS
36th c. B.C. Sumerian pictographs 31st c. B.C. Egyptian hieroglyphs; Sumerian cuneiform 26th c. B.C. Indus valley Harappan hieroglyphs 16th c. B.C. Chinese ideographic script; Semitic syllabary in the Middle East; Hittite hieroglyphs and cuneiform 9th c. B.C. Greeks create first alphabet based on Middle Eastern syllabary; model for later alphabets 7th–5th c. B.C. Latin alphabet derived from Greek 5th c. B.C. Written records kept in China during 2nd half of Chou dynasty 3rd c. B.C. Aśokan Brāhmī script used in India 1st c. A.D. Sogdian script based on Middle Eastern semitic alphabet used in Central Asia	2nd c. A.D. Earliest manuscript remains from Khotan: Prakrit Buddhist text in Kharoṣṭhī script 3rd–10th c. A.D. Mayan civilization uses hieroglyphs 4th c. A.D. Gothic alphabet derived from Greek 6th c. A.D. Earliest manuscript remains in "Tokharian" language used at Kucha and Turfan 6th–7th c. A.D. Sogdian script adapted by Turks for translating Buddhist texts into Turkish 8th c. A.D. Earliest texts in English using Roman alphabet 8th c. A.D. Turkish inscriptions in "Runic" script; Turkish Uighurs adopt Sogdian script 10th–11th c. A.D. Slavic Cyrillic alphabet derived from Greek 11th c. A.D. Buddhist texts translated into Hsi-hsia language	1750 B.C. Hammurabi's code of laws for Babylonia 594 B.C. Solon creates code of laws in Athens 509 B.C. Founding of Roman Republic with constitution 450 B.C. Roman law codified Torah guides Jewish state 320–200 B.C. Confucian codes systematized in China c. 250 B.C. Aśoka promotes Buddhist ethics 439 A.D. Codex Theodosianus: summary of Roman law 7th century Tibetan constitution written 604 A.D. Prince Shōtoku's constitution based on Buddhism and Confucianism in Japan 670 A.D. Laws of Visigoths codified in Spain	802 A.D. Charlemagne codifies Germanic tribal laws 843 A.D. Treaty of Verdun divides Charlemagne's Empire; beginning of European states: Germany, France, and Italy 850–950 A.D. Russian, Magyar (Hungarian), and Polish states founded Written constitutions not common in European states until 18th–19th centuries 1206 Mongols promulgate Imperial Code (Yasa) 1215 A.D. Magna Carta, Bill of Rights, gives nobles right to advise British king 1787 A.D. United States Constitution 1795 and 1875 French Constitution 1812 A.D. Constitution in Spain

CHAPTER THIRTEEN

THE DHARMA COMES TO TIBET

Srong-btsan-sgam-po's most lasting accomplishment was to lay the foundation for the development of the Buddhist tradition in Tibet. Less than a century and a half after the efforts of this king, Tibet became a true center for the Dharma in Asia, a storehouse of sacred texts and a home for Buddhist masters. Thanks to the foresight of Srong-btsan-sgam-po, Tibetan culture was set on a course that would later allow it to make an invaluable contribution to world civilization.

According to the Shes-bya-kun-khyab, Srong-btsan-sgam-po's efforts to bring the Dharma to his country began in his fifteenth year when the young king had a remarkable vision. He beheld an ancient thousand-armed and eleven-faced Avalokiteśvara statue located far to the south in a land where a precious kind of sandalwood known as snake's heart could be found. From this unusual wood, wonderful lifelike statues could be carved.

Knowing that the Tibetan character was so strongly influenced by the ancient demoness ancestry, Srong-btsan-sgam-po felt this statue would be necessary to convert the hearts of the people. He soon sent a mission to India, headed by the monk Ākaramatiśīla, who is said by the Buddhist tradition to have been an emanation of the king.

THE MISSIONS OF ĀKARAMATIŚĪLA

The monk Ākaramatiśīla traveled south, teaching as he went, until he reached the southern tip of India near Śrī Laṅka, where the tīrthika king O-ma-tha Ke-sar ruled. Though all his ancestors had been Buddhists, this king had destroyed temples and encouraged animal sacrifice. To convert the king, Ākaramataśīla flew through the air at a great height, circling round a Kāśyapa stupa near the palace. Feeling faith rise within his heart, the king received teachings from Ākaramatiśīla, who instructed him to build temples and to search out the rare snake's heart sandalwood to make statues.

With the monk's help, the king located a certain beach along the ocean where a herd of elephants was sleeping in the sand. Chasing away the elephants, they found a precious sandalwood tree buried beneath the elephants' resting spot. As they dug it out, light began to pour forth from the tree, and a powerful fragrance of sandalwood spread for twelve miles around. When the trunk was split open, the eleven-faced Avalokiteśvara statue was found inside, vibrant and shimmering, as though alive. Leaving the beach, the monk collected a basket of sand, for he had noticed that each grain was a jeweled Avalokiteśvara throne, surrounded by thousands of ḍākinīs. Returning to the Kāśyapa stupa, he found within it seven chests of Buddha relics, and from three he took a small portion for Tibet.

The monk headed north with the precious objects, stopping at Bodh Gayā where he received leaves and flowers from the Bodhi tree and a piece of the tree itself. He collected more precious sand from eight sacred locations, as well as from the bank of the Nairañjanā river where the Buddha had bathed just before his Enlightenment.

Ākaramatiśīla returned to Tibet and presented the holy objects to Srong-btsan-sgam-po, who rejoiced at the blessings they would bring to his country. Now the king had a vision of another Avalokiteśvara statue, this one located in Nepal in a cedar forest. The monk went forth once more, and deep in the

Nepalese forests, he located a sandalwood tree. Inside he found five Avalokiteśvara statues, which he took back to Tibet for the benefit of the land and the people.

MINISTER MGAR'S MISSION TO NEPAL

The king soon had a third vision of two wonderful statues that could help convert the warlike Tibetans. He saw that one statue was in the possession of the king of China and the other belonged to the king of Nepal. In the royal family of each kingdom was a ḍākinī princess, an incarnation of Tārā. One tradition tells how rays of light from Avalokiteśvara had simultaneously descended into the world at the three royal palaces in Tibet, Nepal, and China, blessing the lineages of these kings, whose children were destined to work for the Dharma. By marrying these princesses, Srong-btsan-sgam-po could bring to Tibet the powerful statues that would ensure the success of the Dharma. Srong-btsan-sgam-po's ministers favored alliances with Nepal and China for political reasons, and strongly supported the king's plan.

Thus, sometime before 624 (see chapter 12), the king sent his chief minister mGar sTong-rtsan-yul-gzung to Aṁśuvarman (r. 576–621?), the king of Nepal, with gifts of precious gems and letters written in Nepalese with gold ink upon blue paper. According to the seventh century Chinese pilgrim Hsüan-tsang, the king of Nepal was a very wealthy and learned man, a scholar of some repute and the author of a grammar. He could trace his lineage back, it was said, for hundreds and hundreds of generations to the era of the former Buddhas.

To this formidable king, mGar brought the Tibetan king's request for his daughter in marriage and for the special Buddha statue as dowry. Though mGar explained to the Nepalese king that Srong-btsan-sgam-po was a great man, in fact an incarnation of Avalokiteśvara, Aṁśuvarman was extremely skeptical. He asked mGar how the king of a wild and barbarous country could consider himself worthy to be the son-in-law of the king of Nepal, who was the scion of an ancient royal lineage and the patron of an established Dharma tradition. mGar should return to Tibet, said Aṁśuvarman, and ask this king of his if he were capable of even having such a fine Buddhist statue carved.

mGar then handed the king the first of the letters from Srong-btsan-sgam-po, all of which are recorded in Tibetan histories. Srong-btsan-sgam-po had written Aṁśuvarman that indeed there was no Dharma in Tibet, and no Dharma treasures, and for that very reason he was requesting the statue. In return, he promised to build 108 temples, all of which would face south toward Nepal.

The Nepalese king was taken aback, suspicious of this strange king who had answered his query many weeks before he had made it. He again asked mGar to find out if his king could establish a system of virtuous laws for the people. In reply mGar gave him the second letter, which informed Aṁśuvarman that the Tibetan king could manifest five thousand emanations, and all of them could write ten different systems of laws.

The uneasy Aṁśuvarman asked mGar how his king dared to compare himself to the wealthy and prosperous king of Nepal. So mGar produced the third letter, which announced that the king of Tibet could easily open up the natural resources of the Tibetan land, a land rich in the gold that all the countries to the south depended upon. If the Nepalese king still hesitated, said the letter, Tibet would send an army 50,000 strong to destroy Nepal and take the princess.

Aṁśuvarman was now convinced that this Tibetan king was indeed a magical and powerful lord, and agreed to send his daughter Bhṛkuṭī and the requested dowry. Crossing the Himalayas, the Nepalese princess soon arrived in Tibet, bringing with her a statue of Akṣobhyavajra (Mi-bskyod-rdo-rje), which is a famous statue of the Buddha at age eight, as well as a statue of Maitreya and a sandalwood Tārā. These rare images would be housed in new temples that would later be erected by Srong-bstan-sgam-po and his queens.

THE ROYAL COURT IS MOVED TO LHA-SA

With the help of his Nepalese wife, known to the Tibetans as Bal-mo-bza' Khri-btsun or Bal-bza', Srong-btsan-sgam-po first erected the white palace of dMar-po-ri in lHa-sa. This shifted the seat of government from the Yar-lung valley, where the ancient kings had ruled, to the north side of the gTsang-po river near its conjunction with the sKyid-chu river.

The new palace contained many inner apartments and was wonderfully decorated on the outside with precious metals. The Ngo-mtshar-rgya-mtsho notes that innumerable turrets and towers rose from the dMar-po-ri, and that the palace was a veritable museum of art and contained a huge library of texts. Yet it was also designed to be a fortress defensible by a handful of men. A broad avenue was constructed leading to the eastern gate, and the king often rode up to the palace on horseback along this road. Many centuries later, in the time of the Fifth Dalai Lama, the Potala palace would be erected on this same hill.

Not far from the dMar-po-ri, the king had discovered the Six Syllable Avalokiteśvara Mantra naturally imprinted in the rock on one side of the river. He had also seen natural images of Tārā, Hayagrīva, and Avalokiteśvara on the other side of the river. Rainbow light radiated in an arc across the river between the forms and the mantras. This marvelous scene was later reproduced by Nepalese artists on the walls of the palace by order of the king. The images were also carved into the rock where they had appeared, and the king himself created other remarkable statues of Vairocana, Amitābha, and other deities.

MINISTER MGAR'S MISSION TO CHINA

After his marriage to the Nepalese princess, Srong-btsan-sgam-po sent the minister mGar to China to request a marriage with the Chinese princess he had seen in his vision. This princess, a daughter of the Emperor T'ai-tsung (the second T'ang

Emperor, r. 627–649), became known to the Tibetan people as 'Un-shing Kong-jo. She is regarded as a great benefactress of the Tibetan culture, and many stories are told about her.

The T'ang annals note that in 634 the first ambassador from Tibet arrived in China, and the emperor sent an envoy to King Srong-btsan-sgam-po. This envoy then returned to China with a Tibetan minister—apparently mGar—requesting a Chinese princess for the Tibetan king. But a T'u-yü-hun envoy arrived in the Chinese court, also asking for the hand of a princess. Srong-btsan-sgam-po gathered his troops and attacked the king of the T'u-yü-hun, forcing him to flee. The Tibetan army 200,000 men strong marched northeast, threatening to invade the Chinese Empire if the Tibetan king were not granted the princess. The emperor sent out 50,000 troops, whose advance guard stole into the Tibetan camp in the night and killed many soldiers. Srong-btsan-sgam-po decided to negotiate once more. He sent mGar with gifts of gold and precious objects to make another request for the princess, and Emperor Tai-tsung consented to consider the request of the Tibetan king.

The emperor was still uncertain about giving a princess to the Tibetans, for better alliances might be made with the wealthy Persians or the powerful Turks. In his court he found no one who supported the Tibetan claim. In order to make an impartial decision, the emperor set up a series of contests among the ministers from each land. The Maṇi bka'-'bum tells how mGar, representing the Tibetan king, went through the many trials and tests of wit, overcoming all obstacles. Buddhist histories explain that mGar was a manifestation of the Bodhisattva Vajrapāṇi.

Impressed with the keen intelligence of the Tibetan minister, the Chinese emperor finally agreed to send the princess to Tibet. The princess, however, hesitated to make her home in a faraway land where the climate was so harsh and the people's customs so different from her own. The Maṇi bka'-'bum recounts how the emperor insisted she go, explaining that the virtue and power of the Tibetan king were evident in his mighty army, his wise envoy, his support of the Dharma. These two rulers were soon to become close allies (see chapter 14).

The emperor then gave the princess many wonderful gifts to take with her to Tibet. To win over the hearts of the Tibetan people she took treasures of jewels and ornaments, rare delicacies, sets of armor and weapons, cushions of silk and fur, and silk clothing. To enrich the culture of her new homeland, she took three hundred and sixty works on divination, treatises on conduct, domestic affairs, magic, medicine, animal husbandry, and other disciplines, as well as dictionaries, glossaries, a Sanskrit manual, and Buddhist texts.

The most important gift from China was a statue of the Buddha at age twelve, a large golden image encrusted with jewels, known as the Jo-bo-chen-po. Many generations earlier, a Chinese emperor had obtained it from an Indian king to help spread the Dharma in China. Buddhist historians note that the Jo-bo-chen-po had been created by the artist Viśvakarman during the lifetime of the Buddha and had been blessed by the Buddha himself. This statue was said to possess the power to hasten the enlightenment of anyone who came into contact with it.

Thus the princess departed on her long journey. The emperor tried to detain the marvelous mGar in his court, even offering him a Chinese princess in marriage, but mGar escaped. Angry at being detained, mGar inflicted considerable damage in China, tricking the emperor into burning great quantities of fine silk, putting many herds of sheep to death, and ruining the harvest.

KONG-JO'S ARRIVAL IN TIBET

In the meantime the Chinese princess was waiting in the eastern borderlands for mGar to escort her to lHa-sa. While she was in Khams, she showed the people how to plant crops in a new way using wild antelope to plow, and how to grind grain with a water-powered wheel. Along a mountainside, she had an eighty-foot-high statue carved in rock. In 'Go-jo and lDan-ma she had verses from a Buddhist text carved in rock. When she reached Sum-pa, mGar finally rejoined her, and they started for lHa-sa.

According to the T'ang annals, Srong-btsan-sgam-po came to Ho-yuan just east of Koko Nor lake to meet the princess. This account describes how the princess left China with a huge escort while the Tibetan king went north at the head of his army. After the marriage, which was presided over by T'ang ministers, the king expressed his delight at marrying the Chinese princess. Since the princess disliked the Tibetan custom of painting the face red on special occasions, the king ordered that this practice be stopped. He also promised to build her a beautiful residence, and according to this account, began to wear royal attire similar to that used in the T'ang court.

The Maṇi bka'-'bum gives a more religious interpretation of events. Once the princess had started toward Tibet, Srong-btsan-sgam-po let it be known that she was an incarnation of a goddess and might arrive from any direction that she pleased. All across Tibet, the mountain roads were repaired, and the people kept watch. Inhabitants of the east said she passed through Mal-dro, while those in the south said she had crossed the slopes of the mountains. The people in the west said she passed through sTod-lung, while those in the north saw her cross the pass at 'Phan-yul. Each of these locations still bears a place name associated with the Chinese princess. Her arrival at the capital was the cause for a great celebration. People put on their finery and ornaments and went cheering through the streets.

FOURTEEN SPECIAL TEMPLES

After she arrived in lHa-sa in the iron-ox year of 641, Kong-jo realized that there were powerful negative forces at work that could prevent the development of the Dharma. She saw that the land of Tibet was like the body of a demon, lying on its back, with its heart located at the little lake of 'O-thang in lHa-sa.

But she also saw the great potential of this new land, and she knew that one day people of faith would come from all directions to Tibet. With her skills of geomancy and divination and with the blessings of Tārā, she could predict where various royal buildings should stand, and where the temples should be built. To convert the heart of the demon at 'O-thang, she brought the Jo-bo-chen-po into lHa-sa. When the wagon carrying it sank into the sand right near 'O-thang, she draped silk around the statue and performed a special ceremony.

With the help of the Chinese princess, Srong-btsan-sgam-po built specially designed temples, known as mtha'-'dul, yang-'dul, and ru-gnon, in precise locations throughout Tibet to immobilize the demon. On one of the demon's shoulders in Mal-dro northeast of lHa-sa he built sKa-tshal. On the other shoulder north of sNe'u-gdong in Yar-lung he erected Khra-'brug. In this temple, Srong-btsan-sgam-po had Nepalese artists portray events from his life in paintings that covered the walls. On the demon's legs in gTsang, the king built gTsang 'Gram-pa and Grom-pa-rgyang, which is near lHa-rtse along the western course of the gTsang-po river. These four temples are known as the four inner subduers.

On the demon's right elbow in Kong-po the king built Kong-po-bu-chu along the Nyang river, while upon the left elbow in lHo-brag he established lHo-brag-khom-mthing. On the demon's knees, the king erected sKa-brag and Bra-dum-rtse. These four temples are known as the border subduers.

The most distant temples were built upon the demon's hands and feet. On one hand in the northern plains he erected Rlung-gnod of Byang-tshal, and upon the other hand in the east in Khams he built Klong-thang-sgron-ma in 'Dan-khog. The feet were located in Mang-yul and Mon-yul. In Mang-yul he built Byams-sprin and in Mon-yul he built Bum-thang sPa-gro-skyer-chu along the Paro river in Bhutan.

The Nepalese queen Khri-btsun had wished to be the first to build a temple, but her efforts had been undone no matter where she tried to build, for the land had not been pacified. The Maṇi bka'-'bum describes her unhappiness and her jealousy of the young Kong-jo. Disheartened at the trouble she seemed to be causing, Kong-jo began to think of returning to China, but mGar convinced her to stay.

To protect her position as the older queen, Khri-btsun apparently prevented the new Chinese queen from living with the king for some years. The Tun-huang documents note that Kong-jo had arrived in 641, but only lived with the king the last three years of his life, from 646 to 649.

Eventually, however, Khri-btsun sent Kong-jo a valuable gift of loose gold, and asked for her help in building temples. Ra-sa-'phrul-snang was then built in lHa-sa in the center of what had been the little lake of 'O-thang. A famous temple in India, at the Buddhist center of Vikramaśīla, was the model for the Ra-sa-'phrul-snang.

The Chinese princess now built the temple of Ra-mo-che where the Jo-bo-chen-po was then housed. The Akṣobhyavajra statue brought by the Nepalese queen was put in the Ra-sa-'phrul-snang gTsug-lag-khang. After the death of the king, the locations of the statues were changed, the Jo-bo-chen-po being moved to the gTsug-lag-khang, and the Akṣobhyavajra, also known as the Jo-bo-chung-ba, being placed in the Ra-mo-che temple. The founding of these two temples was deeply significant to later Dharma Kings, who mentioned them in their edicts and inscriptions.

UNCEASING EFFORTS OF THE KING AND HIS TWO QUEENS

After the founding of temples in Tibet, Srong-btsan-sgam-po sent a minister to China to request permission to build temples at Wu-t'ai-shan, a mountain in northeastern China sacred to Mañjuśrī. To fulfill his promise to the king of Nepal that he would build a large number of temples, Srong-btsan-sgam-po built temples all around this holy mountain. Names of some of these sites can be found in a Chinese history published in 1896. The king established a total of 108 temples in Tibet and China.

Long before, the king had built the nine-storied castle of Pha-bong-kha north of lHa-sa as his meditation retreat. The Shes-bya-kun-khyab describes how the king and his two Buddhist queens now often practiced together there and were given predictions about the future success of the Dharma in Tibet. The rGyal-rabs-gsal-ba'i-me-long notes that temples were also built in the east at Ka-chu, Kan-chu, and Gling-chu. His wife Mong-bza' Khri-lcam built a temple at Brag-yer-pa northeast of lHa-sa, and Thang-skya was built across the river from sKa-tshal not far from the region where Srong-btsan-sgam-po had been born in Mal-dro. There in Mal-dro at Byams-pa Mi-'gyur-gling, five stupas were built, and the castle became famous as a place of powerful blessings. The king himself could often be found there in meditation, and he would greet any visitors who came to offer prayers. Many small hermitages and retreat centers were also established. To erect these buildings, the king employed over three hundred craftsmen and artisans, some of them emanations of the king, according to Buddhist accounts.

Word of the devotion of Srong-btsan-sgam-po and his Buddhist queens reached China, for Buddhist pilgrims from that land now traveled to India via Tibet. The seventh century pilgrim I-tsing notes that Kong-jo herself aided the pilgrim Hsüan-chao, who journeyed to India in 651 A.D. He stopped twice in Tibet, on his way to and from India. The Buddhist works of the king are also mentioned in the Li'i-yul-lung-bstan-pa, which describes Srong-btsan-sgam-po as a great Bodhisattva who became king of Tibet and erected many stupas and monasteries.

Though the Tibetan people considered all their ancient kings godlike beings, the Dharma Kings were recognized as truly rare individuals. An interesting fragment of a document in Tibetan found at Tun-huang praises Srong-bstan-sgam-po and the later king Khri-srong-lde-btsan in this way:

> "They possessed the bodies of men,
> but their ways were those of the gods.
> Among other men, in other kingdoms,
> this has not happened before,
> and is not likely to happen again.
> Such a thing, even among the gods, is rare indeed."

During the lifetime of Srong-btsan-sgam-po, the common people and most of the ministers were not aware of their king's long-range vision. He worked subtly and privately, causing no alarm to ministers devoted to the teachings of the Bon-pos or to people strongly rooted in ancient folk customs. Yet his remarkable intelligence and compassion were widely known and widely appreciated. In the Tun-huang chronicles can be found poetry that praises his reign:

> "On the outside the kingdom reached the four horizons.
> On the inside the royal government held firm.
> All of the subjects were equals, both the great and small.
>
> "Unlawful taxes were now removed,
> and there was leisure enough for all.
> Autumn followed spring in the round of the seasons,
> bringing a good harvest enough for all.
> Desires were satisfied and dangers removed.
>
> "The powerful were restrained, the insolent humbled,
> and the hostile were mastered.
> The sincere were well loved, the wise were well praised,
> and the valiant well honored.
> The law was good, the empire was vast,
> and all of the people were happy!"

BUDDHIST TEMPLES FOR TIBET

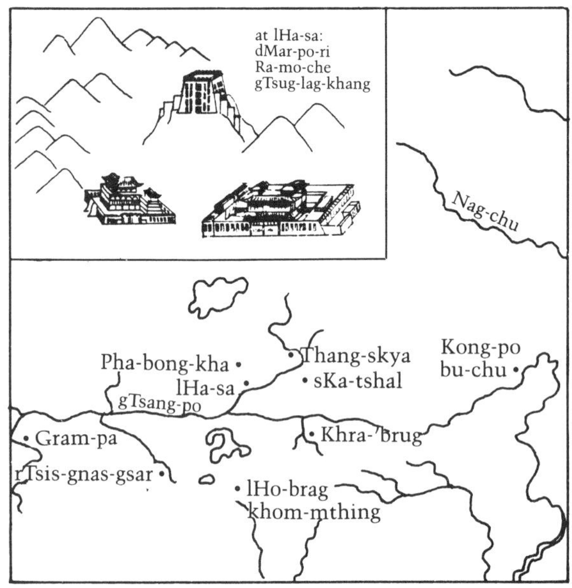

Laying the foundation for the Dharma, Srong-btsan-sgam-po built twelve specially designed temples in chosen locations with the help of the Chinese princess in order to subdue negative forces at work in Tibet. Besides those shown here, Bra-dum-rtse and sKa-brag were built; Klong-thang sGrol-ma was constructed in lDan-khog; in gTsang, Grom-pa-rgyang; in the north, Rlung-gnod; Byams-sprin and sPa-gro-skyer-chu in the south in the region of Bhutan.

MISSIONS TO FOREIGN LANDS

Dharma King Srong-btsan-sgam-po dispatched two of his trusted ministers, Thon-mi Saṁbhoṭa and mGar, on missions of great importance: to establish relations with foreign countries, to initiate cultural exchange, and to obtain precious Buddhist statues for Tibet. mGar negotiated the two marriage alliances for the king with Chinese and Nepalese princesses. A third emissary, Ākaramataśīla, traveled to locations in India and in Nepal to obtain statues and relics.

SEVENTH CENTURY BUDDHISM

The Buddhist teachings spread from India south to Śrī Laṅka, east into Southeast Asia, and north into Central Asia. From there the Dharma eventually reached China, and by the seventh century had passed from China to both Korea and Japan. Seventh century Buddhist pilgrim Hsüan-tsang was aided by several Buddhist Turkish kings. The pilgrim I-tsing was studying Sanskrit in Śrī-vijaya while the kingdoms of Khmer and Nan-chao also had Buddhist Sanghas.

THE SPREAD OF BUDDHISM

Time of Buddha: Dharma flourishes in central India

Time of Patriarchs: Dharma spreads throughout India

Time of Aśoka: Buddhist missions to foreign lands; Dharma reaches Śrī Laṅka

1st c. A.D. Monasteries established in Central Asia

2nd–3rd c. A.D. Translation centers established in China by Central Asian monks

2nd–5th c. A.D. Buddhist centers established in Burma, Cambodia, Vietnam

3rd c. A.D. Khotan flourishes as Buddhist center

4th–5th c. A.D. First Buddhist texts reach Tibet in reign of lHa-tho-tho-ri

4th–5th c. A.D. Buddhist teachers reach Korea

6th c. A.D. Chinese schools of Buddhism develop

6th c. A.D. Western Turks favor Buddhism

6th–7th c. A.D. Japanese monks study in China

7th c. A.D. Foundation laid for Buddhism in Tibet

8th c. A.D. Dharma firmly established in Tibet

9th c. A.D. Uighur Turks convert to Buddhism

10th–11th c. A.D. Hsi-hsia nation becomes Buddhist

13th c. Mongol rulers support Buddhism

16th c. A.D. Mongolia becomes Buddhist

CHAPTER FOURTEEN

THE FOUNDATION OF THE EMPIRE

It was during the reign of Srong-btsan-sgam-po that the tribes on the Tibetan plateau were all united and the foundation was laid for the empire. This king had a great reputation even at age thirteen in the very beginning of his reign. Because of his great compassion and keen intelligence, his physical marks, and his unusual blue eyes, no one doubted that he was an extraordinary individual. His two close advisers and ministers, mGar sTong-rtsan-yul-gzung and Thon-mi Saṁbhoṭa, had both seen the figure of Amitābha upon his head. Tibetan histories mention that though the king wore a red turban, word spread even to the borderlands: The "Two-headed Tibetan King" possessed enormous powers and a huge magical army.

UNITING THE TRIBES

His rule began with the death of his father, gNam-ri, who, according to the Tun-huang documents, had been poisoned. Dags-po, Zhang-zhung, Kong-po, Sum-pa, and Myang-po then revolted. Even though Srong-btsan-sgam-po was young, he searched out the traitors and put down the rebellious clans, who now submitted to him and became his subjects. The minister Myang Mang-po-rje

Zhang-snang received the submission of the Sum-pa tribes in the northern plains of Tibet. According to the Tun-huang chronicles, when the king set forth with his army, before he had even appeared on the road heading north, 'A-zha together with rGya honored him.

The T'ang annals note that some of the Yang-t'ung and other Ch'iang tribes in the west had already come under Srong-btsan-sgam-po's rule sometime before 634. Shortly after 634 the Tang-hsiang, the Pai-lan, and the T'u-yü-hun in the northeast were defeated. These records describe the king's march one hundred miles northeast of rGyal-mo-rong to Sung-pan on the Min river, his defeat of the T'u-yü-hun, and the T'ang emperor's consent to his request for a marriage alliance (see chapter 13).

In addition to his two foreign queens, Srong-btsan-sgam-po married three women from among the important Tibetan families and borderland peoples to form strong alliances with powerful clans. His wives included a noblewoman from Mi-nyag, one from the Ru-yong family, rJe-khri-dkar, and a noblewoman from the Mong, Mong-bza' Khri-mo-mnyen lDong-steng, who was the mother of Srong-btsan-sgam-po's son, Gung-srong. When Gung-srong reached the age of thirteen, his father gave him the throne, and he ruled for five years. This may have been the period when Srong-btsan-sgam-po retired to work on the constitution (see chapter 12). Once he was enthroned as king, Gung-srong married Mang-mo-rje Khri-dkar from 'A-zha, which served to cement ties with 'A-zha. A son Mang-srong was soon born to them; shortly thereafter, Gung-srong died at the age of eighteen, and Srong-btsan-sgam-po stepped back into the government.

COURT INTRIGUES

The more powerful Tibet grew, the more coveted the positions of state ministers became, resulting in continuing court intrigues. Myang Mang-po-rje, who had pacified the Sum-pa for the king, was undermined by Khyung-po Zu-tse, the jealous minister who had served gNam-ri. Zu-tse intimated to minister

Myang that the king no longer trusted him, while suggesting to the king that Myang had been disloyal. In the end, Myang was executed, and Zu-tse grew ever more powerful. But Khyung-po Zu-tse himself then plotted against the king, trying to lure him away from court to visit the regions of rTsang-bod that Zu-tse had subjugated for gNam-ri. Srong-btsan-sgam-po sent mGar to investigate, and when Zu-tse realized he had been found out, he committed suicide. Srong-btsan-sgam-po was not a vengeful king, and when the minister's son approached the king to beg forgiveness for his aged father's misdeeds, he forgave the family.

Disturbed by the scandal, other ministers of the king hastened to renew their oath of allegiance to the king. Srong-btsan-sgam-po renewed all the oaths and even promised to build a tomb for an old retired minister, allowing Bon rites to be performed at the funeral.

Another strange event that suggests continuing intrigue occurred sometime after 641. The Tun-huang documents mention that the king had a younger brother who was betrayed and died in a fire. But little else is known about him.

THE CONQUEST OF ZHANG-ZHUNG

To help stabilize the rapidly expanding empire, Srong-btsan-sgam-po arranged a double marriage alliance in an attempt to peacefully resolve tensions with Zhang-zhung. Tibetan historians agree that Srong-btsan-sgam-po married a princess of Zhang-zhung, Li-thig-sman, and the Tun-huang documents add that he sent his sister, Sad-mar-kar, to be wife of the Zhang-zhung lord, Lig Myi-rhya.

But Sad-mar-kar could tolerate neither the Zhang-zhung customs nor the Zhang-zhung king. Her lament is recorded in the Tun-huang documents in songs full of disappointment and innuendo addressed to her brother in which she urged the king and his followers to "go hunting for wild yak," "to go fishing with harpoons." She sent the king her turquoise ornaments, whose inner message Srong-btsan-sgam-po recognized immediately: If the army could not use turquoise to make the badges of heroes, then they should wear it as jewelry like women! The marriage a failure, Srong-btsan-sgam-po finally invaded Zhang-zhung, about 644, and removed the lord Lig Myi-rhya from power.

After the land of Zhang-zhung had been conquered, a famous exchange took place between mGar and Srong-btsan-sgam-po, revealing the deep bond between them. This speech is recorded in the Tun-huang documents and can also be found in later histories as a citation from old records. It begins in this way:

> "The name of the king is Khri Srong-btsan,
> and the name of the minister is sTong-rtsan-yul-gzung.
> Together, we drink from the river Yar-mo, flowing
> from mDo to rTsang, growing larger from south to north.
> King and minister, together we have conquered
> the dissidents in the four frontiers.
> The king shall never let the people down!
> And never shall the people abandon their king!"

RELATIONS WITH T'ANG CHINA AND NEPAL

Srong-btsan-sgam-po carefully established relations with other ruling dynasties and kings in Asia. T'ang histories of Nepal indicate that after Srong-btsan-sgam-po's father-in-law Aṁśuvarman died, Udayadeva, who had been named crown prince by Aṁśuvarman, was expelled, perhaps killed by one Jiṣṇugupta. Jiṣṇugupta's inscriptions begin in 624, followed by inscriptions of Viṣṇugupta in 640 and 641. Sometime after 641, according to the Tun-huang documents, "Nariba" was made king in Nepal with the support of Tibet. This seems to be Narendradeva, the rightful heir and son of Udayadeva, who, according to the T'ang account on Nepal, was placed on the throne by Tibet and became a protégé of Srong-btsan-sgam-po. Narendradeva's inscriptions in fact begin in the year 641 and extend until 674. Another T'ang history compiled in 650 A.D. also notes that Nepal was a vassal state dependent on Tibet at this time.

The alliance between the Chinese Emperor T'ai-tsung and Srong-btsan-sgam-po was characterized by mutual goodwill and assistance. Once allied by marriage, these two great warriors never fought one another, according to all the old accounts. While Srong-btsan-sgam-po became famous as the founder of the Tibetan Empire, T'ai-tsung, the second T'ang Emperor, became famous as the virtual founder and greatest ruler of the T'ang dynasty. An outstanding military strategist, he defeated the Northern Turks in 630 and extended T'ang dominion west into the oasis states of Central Asia.

T'ang annals note Srong-btsan-sgam-po began cultural exchange with the T'ang court and was very interested in obtaining ancient knowledge traditions for his country. Sons of Tibetan nobles were sent to study in China, and Chinese scholars were hosted in Tibet. Four especially learned Tibetans brought back five major texts on calculation techniques, an ancient science that traced back to Lao-tzu. lHa-lung dPal-gyi-rdo-rje and the Chinese princess Kong-jo helped translate these texts, together with Chinese works on medicine.

THE INVASION OF INDIA

In the earth-monkey year of 648, Srong-btsan-sgam-po came to the aid of the T'ang Empire when the Chinese envoy Wang Hsüan-ts'e was attacked in India. Wang had been dispatched by Emperor T'ai-tsung in 643 and again in 646 to visit the Indian king Harṣa, who had sent an envoy to China in 641. Harṣavardhana, who ruled from 606 until 647, controlled Thāneśwar, Kanauj, and much of northern India. A fervent supporter of Buddhism, Harṣa was a patron of Nālandā and a great admirer of the Chinese Buddhist pilgrim Hsüan-tsang, who visited India between 630 and 644. Military and political relations between Tibet and India are not documented before this event in 648, though the cultural and religious missions are well known (see chapter 11).

The notices on India, Tibet, and Nepal in the T'ang annals explain that the Chinese envoy Wang Hsüan-ts'e did not arrive in

India on his second mission until after the death of Harṣa. Arjuna, the former governor of Tīrabhukti (the region just north of Patna and the Ganges), had already seized northern Bihar. When Wang entered Bihar, Arjuna attacked.

Envoy Wang fled to nearby Nepal where he gained the support not only of Narendradeva but also Srong-btsan-sgam-po, who immediately sent in elite Tibetan troops to protect his father-in-law's envoy. A great battle took place, with 1000 Tibetan troops and 7000 Nepalese cavalry taking 580 walled towns, while the Indian casualties numbered over 10,000 dead, in addition to 12,000 prisoners of war. Arjuna himself was taken captive and sent back to China by Wang.

These accounts give the name of another ruler, the king of Kāmarūpa, Bhāskaravarman, who sent provisions and supplies to the Tibetan, Chinese, and Nepalese troops, and may have been an ally or vassal of Tibet. Bhāskaravarman had been an ally of Harṣa. After the death of Harṣa's enemy, Śaśāṅka (ruler of northwestern Bengal and Magadha until 625), Bhāskaravarman's power in eastern India grew. As described by Hsüan-tsang, Bhāskaravarman was a very learned ruler surrounded by talented men, who came eagerly to his court even from distant lands.

According to recently published Indian histories, after Bhāskaravarman's reign, his three-century-old dynasty was overthrown by a mleccha (foreigner) known as Śalasthambha. Some scholars have suggested that the foreigner might have been a Tibetan, for the term mleccha referred to speakers of non-Indian languages. But the identity of Śalasthambha is not clear.

OTHER FOREIGN CONTACTS

Tibetan histories often summarize Srong-btsan-sgam-po's foreign relations by saying that from China and Mi-nyag he took calculation and astrology texts; from India he obtained Buddhist texts; from Khotan and the Turks he took texts on legal systems; and from Nepal he imported business systems for trade and tapped the natural resources.

There were also contacts with the West. The mKhas-pa'i-dga'-ston and the Shes-bya-kun-khyab mention that during the time that 'Un-shing Kong-jo lived in Tibet, three foreign physicians were invited to Tibet. One came from India, one from China, and one schooled in the Greek medical tradition came from the Byzantine empire. Medical texts from these three traditions were translated, and the Byzantine physician became the court doctor to Srong-btsan-sgam-po. Teaching traditions were founded by each of this physician's three sons.

There were connections with Persia as well. The Deb-ther-dkar-po points out that Srong-btsan-sgam-po adopted Persian court attire — silken headdress, garments of shiny material, and curved slippers — none of which were used in China or India. The Tibetan king also had a strong impact on Khotan, judging from the Li'i-yul-lung-bstan-pa. This text speaks of Srong-btsan-sgam-po as a great Bodhisattva, who inspired the Khotanese kings to continue to protect and support Buddhist monks and monasteries in Central Asia.

Although the Muslims had not yet appeared on the scene, the prosperous Central Asian oasis states were often threatened by nomad invaders from the northern steppes. The Huns had done much damage in previous centuries, and even the T'ang expansion across the Silk Route did not always augur well for these Indo-European Buddhist civilizations. Between 644 and 648, for example, Kucha revolted against T'ang domination, and was almost completely destroyed by Northern Turkish troops in the service of the T'ang emperor. In 648 Khotan became a Chinese protectorate, though its kings remained in power. Just four years earlier, Hsüan-tsang had visited Khotan on his return to China, and described the flourishing state of the monasteries and the learnedness of the people. The king, who is not named, was said to be an extremely devout Buddhist.

It would be very interesting to know more about the relationship between Srong-btsan-sgam-po and these Khotanese kings, who apparently knew him at least by reputation. Bu-ston, for example, mentions that a biography of this Tibetan king was written by two Li-yul monks, perhaps the monks who visited Tibet to study with the king, but the text had been lost by Bu-ston's time.

The king also invited artisans and builders from Khotan, as well as from Nepal and China. The fame of Khotanese artists was widespread across all of Central Asia and China, and several of Khotan's greatest artists lived in Srong-btsan-sgam-po's time. The mKhas-pa'i-dga'-ston mentions that the nine-storied palace of Srong-btsan-sgam-po was built on the pattern of a Sog-po (Saka-Khotanese) castle. Tibetan architecture flourished, and histories note that dwellings were now built more often in the river valleys rather than on the mountainsides.

In the seventh century Tibet became famous for its beautiful statuary, especially intricate images worked in gold. T'ang annals note remarkable gifts from Tibet—golden vessels, wine jugs, and even a whole miniature city crafted in gold. Most of China's imported silver came from the ancient Korean kingdom of Silla and from Tibet. Other exports from Tibet that were sought after by China included yak tails for banners and decoration, marmots, blue-eared pheasants from Koko Nor, a superior kind of honey, salt, musk, various medicines and drugs, and a special cotton dyed a rosy red. Musk was also traded with India; from there it made its way even to the Roman Empire. Tibetan musk was considered the best available and was famous also among the Persians and Arabs.

THE END OF SRONG-BTSAN-SGAM-PO'S REIGN

In 648, the same year he sent troops into India, Srong-btsan-sgam-po sent the Chinese Emperor T'ai-tsung a beautiful seven-foot-high golden statue of a goose to congratulate him on a successful campaign in Korea. When T'ai-tsung died in 649, Srong-btsan-sgam-po sent precious objects to the funeral and had a statue of himself erected at the funeral site. He wrote to the new emperor, Kao-tsung (r. 649–683), promising his continued support and friendship. Kao-tsung sent thousands of pieces of silk and offered the Tibetan king an honorary Chinese title. At the request of Srong-btsan-sgam-po, the emperor also sent silkworms, as well as craftsmen to build windmills, make wine, and manufacture paper and ink.

The following year when Srong-btsan-sgam-po was about to die, he wrote an inscription on a copper plaque, making predictions about the future. He named a king "lDe" to be born five generations later, who would begin to build upon the foundation Srong-btsan-sgam-po had laid. mGar had the plate buried at mChims-phu just before the king died. According to some accounts, when King Srong-btsan-sgam-po passed away, the two Buddhist queens, Khri-btsun and Kong-jo, disappeared together with the king into the eleven-faced Avalokiteśvara statue, but according to the Ka-khol-ma, this was witnessed only by the ministers. A tomb that can still be seen today was built in Yarlung at Don-mkhar near 'Phying-ba. The rGyal-po-bka'-thang mentions that a stone pillar engraved with the history of the ancient kings was set up at the tomb, but it is no longer standing.

SRONG-BTSAN-SGAM-PO'S DATES

The dates of Srong-btsan-sgam-po are still being debated by scholars, for certain records conflict and some events are not clearly dated. The end of Srong-btsan-sgam-po's reign and his death are well established by the events related in the Tun-huang annals and the T'ang annals: the invasion of India in 648, the gifts sent to T'ao-tsung in 648, the letters of 649 between Srong-btsan-sgam-po and Kao-tsung after T'ai-tsung's death in that year; the date of T'ai-tsung's death is also confirmed by Hsüan-tsang. The following year, an iron-dog year of 650, Kao-tsung sent his condolences to Tibet on the death of the btsan-po. The Tun-huang annals also note that the funeral arrangements for King Srong-btsan-sgam-po were begun in the year of the dog, while the Blue Annals and the Deb-dmar concur that the king died in an iron-dog year, which would be 650.

The chronology of the earlier part of Srong-btsan-sgam-po's reign is not so easy to establish. It is generally agreed that Srong-btsan-sgam-po was born in an ox year, but which one is disputed. 'Gos Lo-tsā-ba, author of the Blue Annals, followed the prediction in the Mañjuśrīmūlatantra that Srong-btsan-sgam-po would live some eighty years, and gives his birthdate as 569, an earth-ox year eighty-two years before 650. Bu-ston also states that Srong-btsan-sgam-po lived 82 years and puts the birthdate in a fire-ox year, which would be either 557 or 617, but neither of these dates will fit with a death in 650 to give a lifespan of eighty-two years. Other historians also favor a fire-ox year, but 557 seems most unlikely, for it would make the king nearly one hundred years old at his death in 650, while 617 would give him a liftime of only thirty-three years.

The Tibetan court annals preserved at Tun-huang unfortunately do not offer us a birthdate for this king. Year-by-year entries begin only with the dog year of 650, when the king's funeral arrangements took place. A short summary of earlier events is dated by the number of years before the king's death. The Chinese princess is said to arrive nine years before his death, which would be an iron-ox year of 641, a date also recorded in the T'ang annals and many Tibetan histories. But dates of events before her arrival are less certain because the opening lines of the annals are damaged, and the dates and details are not clear.

Another set of Tun-huang records, known as the chronicles, recounts events in narrative fashion without giving dates. This account shows that after the death of Srong-btsan-sgam-po's father, while the prince was still young, he searched out the murderers and pacified the rebellious tribes. These events were followed by the conquest of the Sum-pa tribes by a minister and the defeat of the 'A-zha by Srong-btsan-sgam-po.

The first dated entry on Tibet in the T'ang annals is 634 when it is noted that Srong-btsan-sgam-po was ruling, that he had become king when he had come of age (Tibetan kings were enthroned in their thirteenth year), and that the Yang-t'ung and other tribes had submitted to him. The subjugation of the T'u-yü-hun ('A-zha) took place between 634 and 641, probably about 636.

To some researchers, these two accounts taken together seem to indicate that the king had just recently come of age when he defeated the 'A-zha, which would make him just a few years over thirteen in 636. Thus 617 would seem to be a better date for his birth, making him nineteen in 636, old enough to take the throne at thirteen in 630, and twenty-four when he married the Chinese princess in 641.

But following this date, we must include a large number of events in thirty-three years, most to be fit into twenty years that he reigned. Some historians such as dGe-'dun-chos-'phel do not find this impossible and cite the amazing accomplishments of Alexander the Great, who lived only thirty-three years.

CHRONOLOGY FOR THE KING'S MARRIAGES AND HEIRS

In addition to the king's numerous accomplishments, another well-known event must be fit into this chronology. Tibetan historians agree that Srong-btsan-sgam-po's son ruled for a period of five years during his father's reign. The son must have been at least thirteen when he took the throne. He died before his father died in 650, and he married an 'A-zha princess, who gave him a child. So we must allow at least fifteen years for Srong-btsan-sgam-po to have a son, Gung-srong, and this son another fifteen years to have a son, Mang-srong. These thirty years will barely fit into the period between 617 and 650.

The necessity of accounting for these two generations helps clarify the chronology of Srong-btsan-sgam-po's five marriages, which is another disputed issue. If we place the marriages to the Nepalese and Chinese princesses before marriages to other wives, then Mong-bza', the mother of Srong-btsan-sgam-po's son, could not have married the king until after 641. But this is clearly impossible, for it would not allow the son Gung-srong enough time to father the grandson Mang-srong.

Some Tibetan sources say that the grandson Mang-srong lived twenty-seven years, making his birthdate 649, for he died in

676 according to Tun-huang records. The T'ang annals note that Mang-srong was too young to rule in 650. But surely the Tun-huang annals would have mentioned his birth if it had occurred anytime after 641, for the births and deaths of other royal princes were carefully noted.

It is also unlikely that the king would have retired and handed over the government to his son anytime after his marriage to Kong-jo in 641. The Chinese princess was given to a ruling monarch, not to a retired father. Thus the son's rule and his grandson's birth must have occurred before 641.

The Blue Annals, on the other hand, says that Mang-srong was thirteen when he took the throne in the iron-dog year 650, lived to be forty-two, and died in the earth-hare year 679. This would put his birth in an earth-dog year of 638.

Other histories also support this earlier birthdate for the grandson, for they note he lived with Srong-btsan-sgam-po a number of years after his father, Gung-srong, had passed away; he was old enough at this time to receive instruction before his grandfather died. Several old historical records place the birth of this grandson as early as 626. In any case it seems Srong-btsan-sgam-po's son could have been born no later than about 625, and perhaps much earlier, so as to marry and have a son before 641. If this is accurate, then the king must have married Mong-bza' sometime before 625.

We can also date the marriage to the Nepalese princess to sometime before 624. We know she married the king during the lifetime of her father, Aṁśuvarman, for mGar dealt directly with him. According to Bu-ston, Aṁśuvarman's era began in 576, which was 242 years before Ral-pa-can, and this date allows an acceptable calculation of the dated inscriptions left by Nepalese kings. Aṁśuvarman was thus in power beginning in 576, apparently ruling jointly with Śivadeva until 606, and alone until 621, the date of his last inscription. He was succeeded by Jiṣṇugupta, whose inscriptions begin in 624. This fits well with the report of the Chinese pilgrim Hsüan-tsang, who noted when he visited Nepal in 637 that Aṁśuvarman had already passed away. Thus it appears that the marriage to the Nepalese princess could not have taken place after the period of 621–624.

HOW OLD WAS SRONG-BTSAN-SGAM-PO?

Based on these dates, a birthdate of 617 is not acceptable. Srong-btsan-sgam-po would not only have had to do the difficult: accomplishing the many works of a mature man between the ages of thirteen and thirty-three; but also the impossible: fathering his son at age eight and marrying the Nepalese princess sometime before his seventh birthday!

The Blue Annals' earth-ox birthday of 569 allows time enough for the king's accomplishments, although it would make the king seventy-two when he married the Chinese princess. It would also make the ministers who served not only Srong-btsan-sgam-po but also his father quite old for active participation in the government—perhaps in their sixties or seventies by the time of their deaths in the 630's and 640's. Yet none of this is as difficult to account for as fathering a child at age eight!

But there would also be a gap of some fifty years between events that are related in the Tun-huang chronicles as though they were a continuous account—the punishment of his father's murderers while the king was young, the conquest of the Sum-pa, followed by the conquest of 'A-zha, which took place after 634.

These difficulties may be overcome, however, if we allow that not all the events of Srong-btsan-sgam-po's reign were necessarily recorded in this particular source. Religious and literary activities, which may well have taken place in the first thirty or forty years of Srong-btsan-sgam-po's reign, are not recorded by the Tun-huang chronicles or annals for any of the Dharma kings. No translations are ever mentioned, for example, nor is the construction of specific temples, even though these activities are fully documented in other sources. Another possibility would be to consider a different ox year as the king's birth—perhaps 605 or 593. But we have been unable to find any source supporting such a choice. Until further documentation is discovered, it seems best to leave the exact birthdate an open question worthy of further research.

THE ERA OF SRONG-BTSAN-SGAM-PO

The seventh century was a period of religious zeal all over Eurasia. The Prophet Mohammed was preaching in Arabia, and Christianity began to spread beyond the Mediterranean and Near East as Nestorian missionaries ventured across Asia. The Buddhist master Dharmakīrti taught in India, and the pilgrim Hsüan-tsang traveled long and dangerous distances to visit the homeland of the Buddha. Even Emperor T'ai-tsung became a devoted patron of the Dharma in the last years of his life.

It was also a time of empire-building. Srong-btsan-sgam-po was contemporary with the first two great T'ang emperors, with the Indian emperor Harṣa, and with Heraclius, who laid the foundation for the Byzantine Empire. Inspired by the teachings of Mohammed, Arab tribes united and began to build an empire.

In his reign, Srong-btsan-sgam-po established an administrative structure for an empire that stretched east past Koko Nor, south into Nepal and India, and included Zhang-zhung in the west. Formal relations were opened with foreign countries, and the Tibetan tribes were united and guided by written codes of law. With the establishment of the first temples and the translation of texts, the teachings of the Buddha began to touch the hearts and the minds of the people. The Tun-huang chronicles close the chapter on this king by noting that the Tibetan people had proclaimed the king "Srong-btsan the Profound" (Srong-btsan-sgam-po), out of gratitude for the benefits he had brought to his land and his people.

TIBET'S NEW INFLUENCE IN ASIA

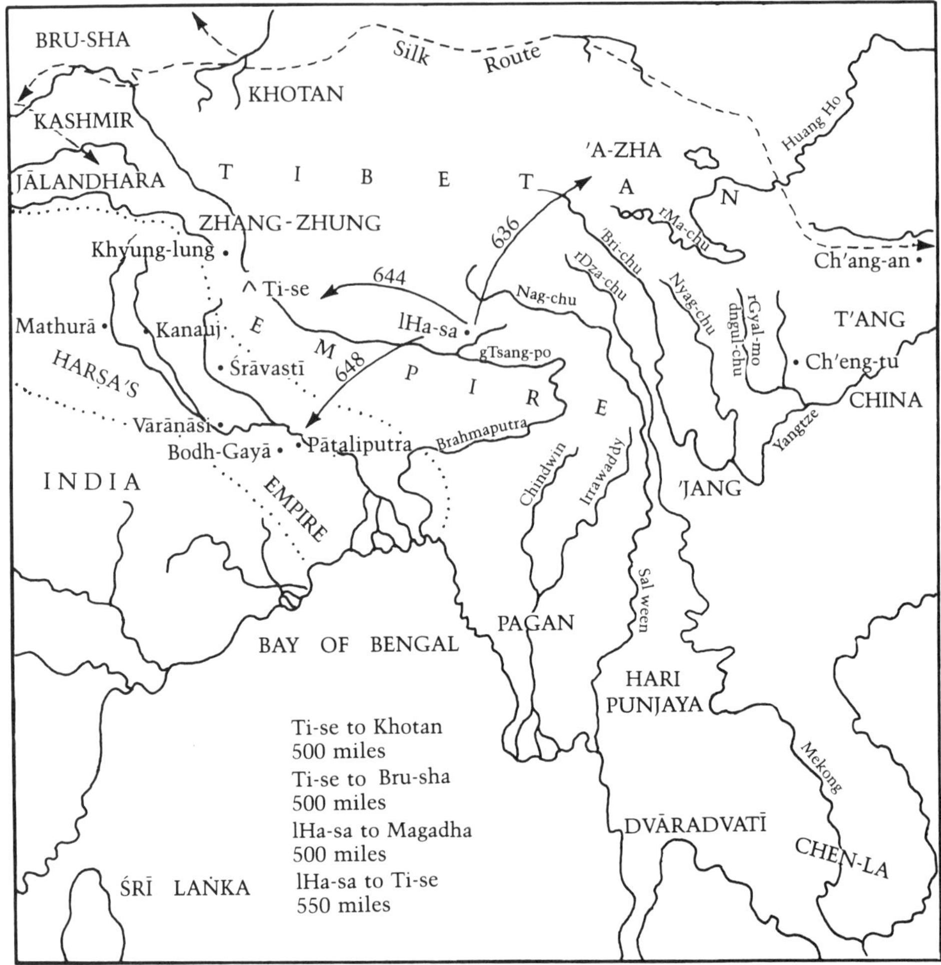

Srong-btsan-sgam-po increased Tibetan influence upon Asian politics by making a marriage alliance with the king of Nepal, and later with the T'ang Emperor. He expanded Tibetan power into the north by conquering and uniting the tribes on the plateau, the T'ang-hsiang, the Pailan, and other Ch'iang tribes, as well as Sum-pa and 'A-zha. To the west he conquered the kingdom of Zhang-zhung, and his influence spread even to Khotan. To the south he invaded India and controlled the throne of Nepal.

THE ERA OF SRONG-BTSAN-SGAM-PO

TIBET	ASIA	EUROPE	WEST
569 A.D. Birth of Srong-btsan-sgam-po in Mal-dro	552 A.D. Turkish Empire founded	c. 550 A.D. Seven Anglo-Saxon kingdoms in Britain	c. 550 A.D. Hopewell Indian culture around Great Lakes in N. America
581 A.D. Srong-btsan-sgam-po ascends throne at age thirteen	r. 576–621 A.D. Aṁśuvarman rules Nepal	552–626 A.D. Juan-juan tribes establish Avar Khanate in E. Europe, S. Russia	Trade established from east coast to Rockies
Thon-mi Saṁbhoṭa develops script and grammar for Tibetan	601–626 A.D. Śaśāṇka rules N.W. Bengal	r. 610–641 A.D. Emperor Heraclius seizes Constantinople from Persians; lays foundation of Byzantine Empire (E. Roman Empire)	Farming common among N. American Indians
Constitution written by king	606–647 A.D. Harṣavardhana rules N. India	611–628 A.D. Persia attempts conquest of E. Roman Empire	Fishing culture on W. coast of N. America grows wealthy
Before 621–624 Marriage to Nepalese princess	618–626 A.D. Kao-tsu, 1st T'ang Emperor	612 A.D. Visigoths regain Spain from E. Roman Empire	c. 600 A.D. Prime of Mayan civilization in Middle America
634 A.D. mGar to China	622 A.D. Hegira of Mohammed	613 A.D. Frankish kingdoms in W. Europe united	c. 600 A.D. Other regional cultures flourish in Middle America: Mount Alban in Oaxaca becomes great city
Frontier tribes submit to king by 634	626 A.D. Durlabhavardhana founds Kārkoṭa dynasty in Kashmir	626 A.D. Avars and Persians ally in failed attempt to take Constantinople	Teotihuacan in Central Mexico grows wealthy; commerce with Mayas
641 A.D. Marriage to Chinese princess	626–649 A.D. T'ai-tsung, 2nd T'ang Emperor	626 A.D. Huns and Slavs free of Avars; Bulgarian khanate founded by Huns	c. 600 A.D. Huari Empire in Peru begins to unite Andes peoples
644 A.D. Conquest of Zhang-zhung	632 A.D. Arab expansion begins across Arabia, Mesopotamia, and Persia		Tiahuanaco Empire in Bolivia influences Huari culture
648 A.D. Tibetan troops sent to India	643–681 A.D. Supported by Tibet, Narendradeva rules Nepal		
649 A.D. Srong-btsan-sgam-po writes to Kao-tsung	645 A.D. Fujiwara's reforms in Japan, modeled on Chinese lines		
650 A.D. Death of Srong-bstan-sgam-po			

CHAPTER FIFTEEN

EXPANSION OF THE TIBETAN EMPIRE

After the death of Srong-btsan-sgam-po, his grandson Mang-srong-mang-btsan became king with mGar as chief minister. Mang-srong had spent the years of his childhood close to his grandfather, receiving personal religious teaching as well as political and practical advice. He apparently learned these lessons well, for under this king, the Tibetan Empire was consolidated and began expanding rapidly beyond the plateau.

The regions that had submitted to Srong-btsan-sgam-po, such as Zhang-zhung and 'A-zha, were now formally incorporated into the empire. In 653 a Tibetan commissioner was appointed to Zhang-zhung, and the administration of this region was organized in 662. Although he died in 667, mGar spent most of his last seven years in 'A-zha, presumably at work on the organization of this new territory. The pacification of 'A-zha was hampered by the machinations of the former ruler, who was not resigned to his loss of independence. The T'ang annals note that Mu-yung No-ho-po of 'A-zha fled to Liang-chou in 670 and requested help in repelling the Tibetans. But the 100,000 T'ang troops sent to aid him were defeated by the Tibetan army. This debacle ended the career of the chief commander of the T'ang forces. He was the first in a whole series of commanders to be demoted for failure against the Tibetans, according to T'ang records.

RELATIONS WITH T'ANG CHINA AND THE TURKS

Some Tibetan sources state that after the death of mGar, T'ang armies threatened Tibet, even reaching lHa-sa and burning the dMar-po-ri palace. We have not been able to find mention of these events in the Tun-huang documents or the T'ang annals, but clearly the relations between T'ang China and Tibet were now growing more hostile. Not only had the T'ang forces aided 'A-zha against Tibet, but the T'ang annals note that once 'A-zha was conquered, there were constant struggles in the northeastern frontiers between Tibetan and T'ang troops.

Records from Tun-huang and the T'ang annals show that the friendly relationship established by Srong-btsan-sgam-po and T'ai-tsung began to change during Mang-srong's reign. His policies were apparently different from his grandfather's, and he began expanding into T'ang-controlled territory. After the pacification of 'A-zha in the east and Zhang-zhung in the west, Mang-srong's hold on the Tibetan plateau was firm. But the lands beyond the Kun-lun mountains, the prosperous oasis states along the Silk Route, were already under the control of the T'ang Empire. By 658 China not only had control of Khotan and Kucha, but had also established protectorates in the far west in Sogdia and Kashmir. The T'ang emperor had also received the submission of the Northern Turks, whose headquarters were along the Orkhon river in the Mongolian steppe.

Sometime before 662 Mang-srong made an alliance with T'ang's adversary the Western Turks. Their homelands lay toward the north around Issyk Kul lake (in modern Soviet Uzbekistan). But by the beginning of the seventh century, the Western Turks had become the overlords of much of western Central Asia, and controlled Bactria, Sogdia, and Tokharia. It had been Turkish khagans who had helped the Buddhist pilgrim Hsüan-tsang cross safely into India, and some of the Western Turkish tribes were Buddhists.

As T'ang power in Central Asia increased, the Western Turks lost ground, and different tribes such as the Qarluq, the Tu-lu, and the Nu-shih-pi formed separate factions. The Tibetan-Turkish alliance was nonetheless strong enough to threaten T'ang control of the Silk Route. An eleventh century Chinese history notes that beginning in 662 combined Tibetan and Turkish forces raided T'ang protectorates. There were attacks on Khotan in 665 and Kashgar in 663. In 667 the Nu-shih-pi submitted to Tibet, and a small country in the far west, known as Wakhan, was also in Tibetan hands.

THE TIBETAN EMPIRE EXPANDS INTO CENTRAL ASIA

According to T'ang sources, by 670 Tibet had taken control of the Silk Route, capturing the four T'ang garrisons at Kucha, Khotan, Karashahr, and Kashgar. The Silk Route was a great prize, for along these caravan routes Chinese silk and Indian ivory, spices, and gems flowed west, while bronze and glassware from the Byzantine Empire traveled east. Kucha was famous for gold, felt, and perfumes, Khotan for jade and carpets, Ferghana for horses and armor. The wars for these crucial caravan routes sometimes caused great damage to Buddhist shrines and temples and also began to affect trade adversely. Travel became risky, for caravans might find themselves caught between two hostile armies.

Tibet's empire had become very extensive by the end of Mang-srong's reign. The old tribal territories of the Yang-t'ung, the Tang-hsiang, and various Ch'iang tribes, who had submitted to Srong-btsan-sgam-po, now were completely annexed. On the east the empire reached the Min river east of rGyal-mo-rong, and in the south it bordered on India, with Nepal a vassal state. To the north and west Tibet held the four garrisons of the Silk Route, bringing their borders all the way to the Turkish lands. Never before had such an empire been seen in the west, said the T'ang historians. They described Tibetan territory as extending more than 10,000 li (3000 miles), reaching "the land of the p'o-lo-men" (brahmins) in the south and the prefectures of Liang-chou Sung-chou, Mao-chou, and Sui-chou (along the Min river) in the eastern borderlands.

The Tibetan kings traveled regularly around the country, setting up summer and winter residences at different locations, as noted in the yearly entries of the Tun-huang annals. This movable court consisted of the king's residence—a felt tent large enough to hold over one hundred people—together with the smaller tents of the ministers, which were set up around the king's. Beginning in Mang-srong's reign, councils convened by the chief ministers or the king were held in various parts of the country twice a year. The ministers swore an oath to the king at a yearly ceremony, and once every three years took a great oath.

YOUNG 'DUS-SRONG BECOMES KING

After ruling over twenty-five years, Mang-srong died in the fire-rat year of 676. Some sources mention that his death was kept secret for three years, which would explain why the T'ang records give the year of his death as 679. His son 'Dus-srong Mang-po-rje had been born to Queen 'Bro-bza' Khri-ma-lod not too long before his father died, but the Tun-huang records do not give a date. T'ang annals note that 'Dus-srong was eight years old in 679. The young child was enthroned with mGar's second son, Khri-'bring, as regent. Several years later Kong-jo, the Chinese queen of Srong-btsan-sgam-po, died in the iron-dragon year of 680. She had lived in Tibet nearly forty years. The T'ang court sent condolences, just as they had on the death of Mang-srong.

Since 'Dus-srong was still a minor, his ministers took charge of foreign relations. They led a number of military expeditions, expanding Tibetan influence to both the southeast and northwest. T'ang annals show that in 678 Tibetan troops seized a crucial T'ang garrison at An-jung along the Min river north of Ch'eng-tu. They would hold it over sixty years as a borderland outpost. That same year another T'ang commander was demoted for lack of success against Tibetan forces. According to an eleventh century Chinese history, in 678 the Erh-ho people around Lake Ta-li in 'Jang submitted to Tibet and asked for an alliance against the T'ang empire.

Tibetan troops were also frequently reported in the lands to the north and west. Tun-huang annals note that in 675 and 676 a Tibetan minister had gone to lTang-yo in Dru-gu-yul, the Western Turkish lands. These visits were apparently connected to efforts by the T'ang Empire to regain the Silk Route. The T'ang annals remark in an entry of 676 that a battle was waged with the Tibetan-Turkish alliance for the four garrisons along the Silk Route. Again between 687 and 689 the minister mGar Khri-'bring led troops to Dru-gu Gu-zan, which may be the Kucha region or perhaps old Kuṣāṇa lands.

EXTENT OF THE TIBETAN EMPIRE

Exactly how far west beyond Kucha and Khotan Tibetan troops reached is not clear, for lTang-yo and Gu-zan have not been identified with certainty. It seems possible that Tibetan armies entered Sogdia, for the Tun-huang annals note that in 694 the Tibetan minister mGar sTag-bu was captured by Sogdians. But nothing else is known of this event, which may have taken place in Sogdia or in one of the many Sogdian merchant outposts along the Silk Route.

Sogdia lay to the west of Khotan and was inhabited by peoples who spoke an Indo-European language related to Khotanese. As early as 627 the Sogdians had begun building the four great cities of Lop Nor north of Tibet between Tun-huang and Khotan. Eventually they spread east into Ordos under the large loop of the Huang Ho east of Koko Nor. Later Sogdian inscriptions, probably from the ninth century, have been found at Drang-tse near Pang-gong, and Sogdian texts have been found at Tun-huang. But the extent of the contact between Sogdia and Tibet will require further research.

In 689 T'ang forces attempted an invasion of Tibet, but had no success, and the commander was demoted, according to the T'ang annals. Between 692 and 693 however, the T'ang armies regained control of the Silk Route and the Tarim basin. They held this region for a century before Tibet reconquered it.

Though they had temporarily relinquished some of their holdings along the Silk Route, the Tibetan troops had great confidence in their military abilities. In the Tun-huang chronicles can be found a lively exchange between the Tibetan general mGar Khri-'bring and the T'ang commander in the borderlands, one Wong-ker zhang-she (Wang Hsiao-chieh), who had recaptured the Silk Route oasis states. The commander sent the Tibetan general sacks filled with grain, announcing that he would have as many soldiers ready to fight the Tibetans. But Khri-'bring replied that great numbers mattered little — many small birds made a meal for a single eagle. The bTsan-po of Tibet, he continued, is like the sun, the Emperor of China is like the moon. Although they are both very great kings, their brilliance in the sky is quite different.

The next battle with the T'ang troops took place in 695–696 at a place known as Su-lo-han in the northeastern borderlands, between Lin-tao and Liang-chou. Khri-'bring and his brother bTsan-ba defeated T'ang troops so badly that the battlefield became known as the Chinese Cemetery at Tiger Pass. Wang Hsiao-chieh became the fourth T'ang commander to be stripped of his rank.

In 696 mGar Khri-'bring went to Ch'ang-an to try to arrange peace. The T'ang annals record his bold offer, which was contingent upon T'ang China removing all troops from Central Asia and dividing the Western Turk tribes equally between T'ang and Tibet. The Empress Wu (r. 684–705) refused to negotiate.

The losses in Central Asia did not end the Tibetan-Turkish alliance. Tibetan troops moved into the far west into Turkish territory beyond the four garrisons of the Silk Route. Already in 694 when 'Dus-srong was about eighteen, Khagan Ton-ya-bgo, the Turkish prince known in T'ang records as A-shih-na T'ui-tzu, chief of the Western Turkish Tu-lu tribes in Dzungaria, had visited the Tibetan court. By 699 the Tu-lu tribes established a firm alliance with Tibet, and the Tun-huang annals note that the khagan personally led Tibetan troops west near his own Turkish domains. The westward reach of the Tibetan Empire is clear, for T'ang records on the Turks mention that in 700 Tibetan troops under this Turkish prince were in Ferghana, five hundred miles northwest of Khotan and fifteen hundred miles from lHa-sa. The ninth century Arab historian Ṭabarī also reports that in 704 Tibetan troops were in Tokharia, which lay west of Gandhāra and Oḍḍiyāna near territories newly acquired by the Arabs.

THE RIGHTFUL PLACE OF THE KING

Though 'Dus-srong was now about twenty years old and personally held the reins of the government, his ministers had had the whole of his minority, a time of many great military victories, to increase their power and prestige. Between them the mGar brothers commanded the entire Tibetan army, while mGar Khri-'bring was chief minister in central Tibet. mGar bTsan-ba took charge of the eastern borderlands with the T'ang Empire, struggling against T'ang troops in battle after battle for thirty years. But the young king may not have authorized or even known of all of his expeditions.

When 'Dus-srong realized in 699 that the ministers of the mGar family were operating as independent warlords, he moved to destroy their power. While Khri-'bring was absent from central Tibet, the king pretended to be organizing a great hunt and collected his most trusted warriors, who turned on the mGar family's supporters and crushed them. The king then sent for Khri-'bring and his brothers, who refused his command to appear in lHa-sa. So 'Dus-srong personally marched north with his troops. Confronted by the king, Khri-'bring surrendered without a fight. The T'ang annals note that mGar Khri-'bring later committed suicide, and bTsan-ba fled to T'ang China, which strangely enough, gave its scourge and opponent great honors. According to Tun-huang documents, 'Dus-srong sent an official letter of friendship to the T'ang government at this time, and peace envoys traveled back and forth between lHa-sa and Ch'ang-an.

After 'Dus-srong's victory, he asserted the rightful place of the king above his ministers in a speech recorded in the Tun-huang documents. This speech is very different from the pledges

King Srong-btsan-sgam-po and Minister mGar sTong-rtsan-yul-gzung had exchanged fifty years earlier:

> "In olden times, in the days of the beginning . . .
> the blue heavens on high never fell,
> the earth below never reached the sky.
> But now I see upon the ground
> worms and ants trying to imitate birds!
> However much they desire to fly,
> they have no wings for soaring.
> Even if they sprouted wings,
> the sky is much too high for them—
> they would never get past the clouds.
> Not reaching the sky, yet not firmly on the ground,
> these dreamers fit nowhere
> and soon become the prey of the hawk . . .
> Tomorrow and tomorrow and tomorrow,
> the ministers will never rule the king—
> the horse will never ride the horseman."

'DUS-SRONG THE WARRIOR KING

'Dus-srong is always described as an immensely powerful individual, both physically and mentally, able to wrestle tigers and kill wild yak bare-handed. He was said to be different from other men, and people in Tibet and even the neighboring kings called him 'Phrul-gyi-rgyal-po, the Magical King. Very tough-minded in military matters, he often led Tibetan troops into battle himself. T'ang annals note, for example, that in 702 the Tibetan btsan-po personally led 10,000 troops into four battles in the northeast. An old Tibetan account notes that 'Dus-srong destroyed the forts and castles of any neighboring state that was not friendly to Tibet and imposed tribute, but unfortunately does not name specific regions.

We do know that in addition to the conquests in the far northwest 'Dus-srong received in 703 the submission of a group of the White and Black Mya-ba, tribes who lived in the region of modern Yunnan. This land was known in the seventh and eighth centuries as Nan-chao in T'ang histories and 'Jang/lJang in Tibetan records. Chinese sources on Yunnan history mention Ai-lao, I-jen, Man-tze, and Chao-chen tribes that came under Tibetan control about this time, and these may be the Mya-ba. Important trade routes ran through the Yunnan regions into Burma, and this region was becoming another zone contested between the T'ang Empire and the Tibetan Empire.

Part of 'Dus-srong's foreign policy involved marriage alliances. He apparently married a Turkish princess, for the Tun-huang annals note that Ga-tun, the Turkish bride, had died in 708. To strengthen bonds with the 'A-zha territory, the Tibetan princess Khri-bangs was married to an 'A-zha chief in 689 A.D. and had a son known as Ma-ga Thogon Khagan.'Dus-srong himself married two women from important Tibetan clans, 'Dam-gyi Cog-ro-bza' and mChims-bza' bTsan-ma Thog-thog-steng.

Despite 'Dus-srong's reputation as a warrior, he supported the Buddhist teachings that his great grandfather had established. He built a temple known as Gling-gi-khri-rtse, which may have been near Khri-rtse, one of the king's winter residences noted several times in the Tun-huang annals. This temple is ascribed to 'Dus-srong in the inscription at sKar-chung written by King Sad-na-legs a century later. It is not certain if this Khri-rtse is the same Gling-chu Khri-rtse named in the rGyal-rabs-gsal-ba'i-melong as a temple founded by Srong-btsan-sgam-po in the east. Bu-ston also names a Ling-chu-khri-rtse as a temple founded by 'Dus-srong's son Mes-ag-tshoms. It may be impossible to determine the exact identity of this temple, for old temples sometimes were refurbished and enlarged, and "refounded" in later times.

According to Tun-huang records, 'Dus-srong, the warrior king, died in battle in the wood-dragon year 704 during an expedition to Mya-ba territory. Soon after his death there was a revolt in Nepal and Se-rib, a region southwest of Tibet, which seems to be in the Kali Gandaki region of modern Nepal. The Se-rib king was captured in 709, and the revolt apparently was put down because Tibetan kings continued to spend summers in Nepal. The T'ang annals also report that Nepal and "the land of the p'o-lo-men" (brahmins), both kingdoms tributary to Tibet on its southern frontier, had revolted. Exactly how far Tibetan control extended into India is not certain. It seems that after the 648 invasion, some territory must have been annexed. According to T'ang records, the btsan-po had gone to Nepal because of the revolt and met his death there in 705 in battle.

'Dus-srong was buried in the royal burial grounds near Yarlung, his tomb next to that of his father, Mang-srong. These two kings, the grandson and great-grandson of Srong-btsan-sgam-po, greatly increased Tibetan military presence in Asia. The Silk Route had been won and lost, but the Tibetan policy of expansion was an established fact. Now her neighbors had to decide whether to become her allies or her opponents.

SEVENTH CENTURY OASIS STATES

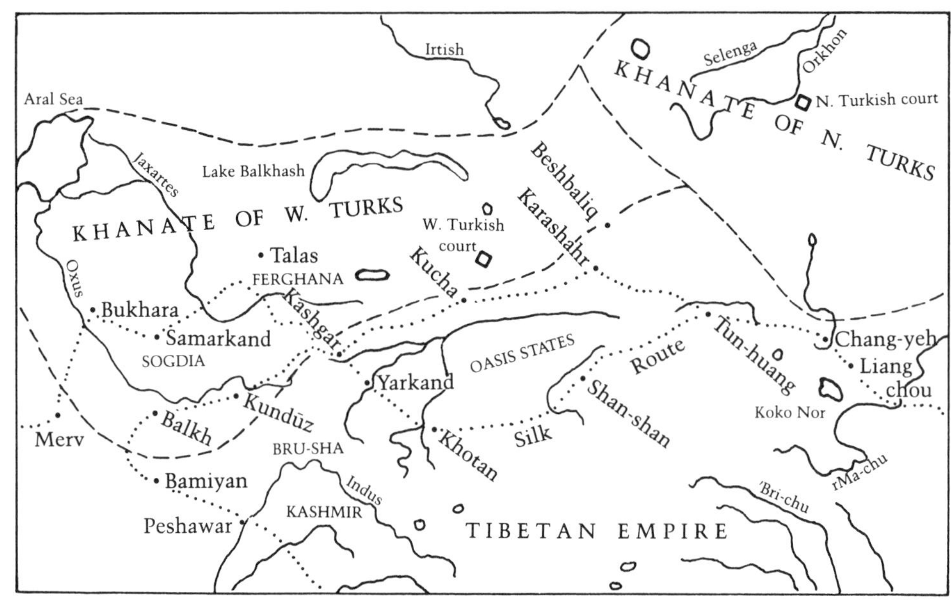

HISTORY OF THE OASIS STATES

750 B.C.
Iranian Scyths, Sakas in W. Central Asia

c. 250 B.C.
Khotan established, Kucha established

210–48 B.C.
Hsiung-nu into Central Asia

127 B.C.
Han dynasty forts at Liang-chou, Kan-chou, Tun-huang

112 B.C.
Silk Route opens

c. 100 A.D.
Kuṣāṇas control W. Central Asia, Chinese power wanes

350 A.D.
Juan-juan from Mongolia invade oasis states

425 A.D.
Hephthalites rule Sogdia, Kabul, Gandhāra, Tokharia

445 A.D.
T'u-yü-hun control S.E. oasis states

552 A.D.
Turks establish empire, occupy land west of Oxus by 563

618–658 A.D.
T'ang dynasty asserts control:
618: Kucha
630: defeats N. Turks
632: Kashgar, Khotan
635: Yarkand
640: Turfan
648: Karashahr

683 A.D.
Restoration of Turkish power

THE ERA OF MANG-SRONG AND 'DUS-SRONG

TIBET	ASIA	WEST	OTHER
650–676 A.D. Mang-srong reigns in Tibet	645 A.D. Downfall of the Soga in Japan	c. 460–711 A.D. Visigoths rule Spain	c. 650 A.D. Huari Empire and Tiahuanaco Empire absorb smaller Andes coastal states
662 Turkish-Tibetan alliance begins	Japan begins period of borrowing from China	641 A.D. Arabs conquer Egypt	Farming tribes in central S. America
667 A.D. Death of Minister mGar	649–683 A.D. Kao-tsung rules T'ang China	661–750 A.D. Under Umayyad Caliphs, Jews and Christians not persecuted	Most N. American chiefdoms are agricultural; hunters on the central plains; fishermen on the west coast
670–692 A.D. Tibet controls Silk Route; empire reaches to Min river in east, to Turkish lands in north, to India in south	659 A.D. Some W. Turks migrate to eastern Europe	661 A.D. Caliph Ali, nephew of Mohammed, is murdered	c. 700 A.D. Growth of kingdom of Ghana in W. Africa, stimulated by trans-Sahara trade for African gold
676–704 A.D. 'Dus-srong reigns in Tibet	663–668 A.D. Silla conquers Paekche and Korguryŏ, unites Korea	672–735 A.D. Venerable Bede, English monk and historian	c. 700 A.D. Strong Mexican influence reaches S.W. and S.E. N. America
687–89 A.D. Tibetan troops in Turkish lands	674 A.D. Arabs reach Indus river	680 A.D. Asiatic Bulgars invade Balkans	c. 700 A.D. First Pueblos built in N. America
692–694 A.D. China regains Silk Route oasis states	683 A.D. Arabs reach Oxus river	686 A.D. All English kingdoms now Christian	c. 720 A.D. Iron and cereal farming spread south of tropical rainforest belt in S. Africa
699 A.D. mGar brothers disgraced	683–734 A.D. Turkish power restored under A-shih-na clan	687 A.D. Frankish kingdom dominated by Carolingians	750 A.D. Teotihuacan, great Mayan city in Mexico, is destroyed and abandoned
700–704 A.D. Tibetan troops in Ferghana and Tokharia	685 A.D. T'ang Empress Wu supports Buddhism	700 A.D. Arabs conquer Algiers in north Africa	
703 A.D. Black and White Mya-ba in 'Jang submit to Tibet	694–704 A.D. A-shih-na T'ui-tzu Turkish ally of Tibet	c. 700 A.D. Independent Slavic kingdoms: Croatia, Serbia in S.E. Europe; Slavs invade Greece	
	700–740 A.D. Yaśovarman rules Kanauj; defeated by Lalitāditya of Kashmir		

CHAPTER SIXTEEN

THE REIGN OF MES-AG-TSHOMS

Khri-lde-gtsug-btsan, commonly known as Mes-ag-tshoms, was the son of 'Dus-srong Mang-po-rje and his queen mChims-bza' bTsan-ma Thog-thog-steng. He was born in 704, the same wood-dragon year of his father's death. It seems that 'Dus-srong's sons struggled over the succession, for the Tun-huang annals especially emphasize that Khri-lde-gtsug-btsan was the one to be crowned king. The T'ang annals remark that a long dispute among the princes followed 'Dus-srong's death. Whether this was a simple power struggle or whether important differences in policies were at stake is not known.

Of the other princes, the brothers or perhaps the cousins of the new king, we know almost nothing, except for one obscure remark in Tun-huang annals that an older brother was deposed in 705. The Tun-huang annals do not explain how he came to be ruling or why he was deposed, but his disgrace may have been connected to the revolt of the southern territories, which happened about this time.

In 706 Gung-srong's wife, the 'A-zha princess Mang-mo-rje, died. The Queen Mother Khri-ma-lod, Mang-srong's wife and 'Dus-srong's mother, became very powerful during the early years of her grandson's reign. Until her death in 712, when he was formally enthroned as king, Mes-ag-tshoms ruled only as Prince.

PRINCESS KIM-SHENG KONG-JO

Six years after his reign began, Khri-lde-gtsug-btsan married Kim-sheng Kong-jo, a Chinese princess belonging to the family of the emperor Chung-tsung. Mes-ag-tshoms' son, lJang-tsha lHa-dbon, was supposedly the intended husband of the Chinese princess, but since the young man died before the princess arrived, she married the elderly Mes-ag-tshoms instead.

There seems to be some disagreement over the age of the king upon his marriage to the princess, for the dates given in the Tun-huang documents and the T'ang annals do not fit well with the commonly accepted account given in the sBa-bzhed. The T'ang and Tun-huang records state that the princess arrived in Tibet in 710. In this year Mes-ag-tshoms would have been but six years old according to Tun-huang records, or twelve years old, according to T'ang records. The Blue Annals, which says she arrived in 712, gives no birthdate for Mes-ag-tshoms. Other histories, however, cite letters from the king to the princess that explain his son's death and leave the decision to come to Tibet up to her.

Nonetheless, all sources agree that the king married Princess Kim-sheng, who was the daughter of the Prince of Yong, the grandson of Emperor Kao-tsung (r. 649–683) and Empress Wu (r. 684–705), the parents of the reigning emperor, Chung-tsung (r. 705–710). Though she belonged to the family of the emperor's brother, she had been raised by the emperor as though she were his own daughter, and he was very fond of her. Recalling the matrimonial alliance between Srong-btsan-sgam-po and T'ai-tsung, the emperor agreed to part with his beloved Kim-sheng in the interests of peace and friendship.

Accompanying her to Shih-p'eng in the borderlands to meet the Tibetan envoys, the emperor changed the name of the town to "Sad Departure." As a wedding gift, he gave Tibet a fertile region southeast of Zi-ling (Xining) known as the Nine Loops of the Huang Ho. The T'ang annals also note that the Tibetans had originally accepted the Huang Ho as boundary, but now the river was bridged, and two walled cities were built on the other bank.

RELATIONS WITH T'ANG CHINA

The bridge and encampments were soon destroyed in battles, for despite the marriage alliance, fighting between Tibet and T'ang forces did not cease. Political and diplomatic relations between lHa-sa and Ch'ang-an were strained at the beginning of the reign of the new emperor Hsüan-tsung (r. 712–756). In 714 Mes-ag-tshoms sent messengers to Hsüan-tsung, recommending that they renew the alliance made with Emperor Chung-tsung. Princess Kim-sheng also wrote the emperor to encourage friendship between her husband and the new emperor. But Hsüan-tsung was not interested in peace talks.

The T'ang annals explain that the Tibetan diplomatic dispatches were elegantly and proudly worded and that their envoys insisted on the rites and ceremonies of an equal state. Hsüan-tsung was also concerned about the military situation, for the T'ang annals note he had studied the maps and was giving his recommendations to T'ang commanders about the strengths and weaknesses of the various troop positions. Peace seemed completely out of the question.

T'ang and Tibetan armies had been at war now for some years. T'ang annals note that Tibetan soldiers had invaded Lin-tao along the Klu-chu (Tao river) as well as Lan-chou, Wei-chou, and other districts in the northeast, and hard fighting was taking place along the frontiers. Emperor Hsüan-tsung was planning personally to lead a huge command of troops into Tibet for a surprise attack. But T'ang officials pointed out that the situation was bad—they had not been able to recover any of the lost territories. Some advisers to the emperor recommended making a treaty, for the T'ang troops were wearing themselves out year after year, and the whole region was being devastated. One official, hoping for peace, suggested that the dispatches that so angered the emperor might have been written by some Tibetan general rather than the btsan-po. Others advised raids against Tibet and espionage. Nearly fifteen years passed before a treaty was concluded, despite successive missions back and forth between the two rulers.

Tibet again dispatched envoys to Ch'ang-an in 729. Finally Hsüan-tsung sent an envoy to Tibet to discuss the possibility of a treaty. Mes-ag-tshoms brought forth all the past correspondence over several years, which he discussed in great detail with the envoy. He then composed a new letter to the emperor, which is recorded in the T'ang annals.

Mes-ag-tshoms pointed out that the military was making peace efforts difficult—border generals on both sides had a way of making trouble. Lovers of battle, they had waylaid several of his peace envoys already. As a younger man, the king said, he had sometimes been swayed by the military commanders. But he truly desired peace. The king wrote that he was well aware of the father-in-law and son-in-law relationship that had been established long before. Because Mes-ag-tshoms himself had married a T'ang princess, the relationship between himself and the emperor was similar to being in the same family. Since this brought peace and joy to all peoples, the Tibetan king promised never to be first to violate a treaty of alliance, if the emperor wished to renew the old friendship.

In 730 a peace treaty was finally signed, establishing the border east of Koko Nor in the Red Hills at the pass known as Chih-ling (tentatively identified by Demiéville as mDo-nyi-zla, the Mountains of the Sun and Moon). In 733 Mes-ag-tshoms wrote again to Emperor Hsüan-tsung. T'ang and Tibet were equally great kingdoms, said Mes-ag-tshoms, and he hoped that the peace would endure. He expressed a desire to have the treaty inscribed on a border pillar so the peoples in the northeast would cease hostilities. A marker engraved with the treaty was erected in 734, though it reportedly was torn down shortly afterwards. But envoys did travel regularly between the two capitals for the next fifteen years. After recording the treaty and the letter, the T'ang annals mention in passing that long before this treaty a wall had been built along the borderland with guards posted at intervals, but the exact location of the old wall is not given in the T'ang records.

About this time the Chinese princess Kim-sheng asked that the five Chinese classics be sent to Tibet, and these were copied and dispatched, though not without some misgivings. Texts on war strategy were among them, and the emperor's advisers feared the Tibetans would use them to improve their military tactics. One official remarked that the Tibetans were extremely resourceful and quick to learn, and he feared disaster.

NEW ALLIANCES ARRANGED BY MES-AG-TSHOMS

It was no wonder that the T'ang officials were uneasy. Though T'ang forces had captured the Silk Route, they now faced a triple alliance engineered by Tibet with the Turks and the Arabs. Mes-ag-tshoms also strengthened Tibet's alliances with other countries, for the Tun-huang documents show envoys arriving from the Türgish tribes of the Western Turks, from Nan-chao ('Jang), as well as from the Arabs.

Under the leadership of the famous general Qutaiba, the Arabs had begun invading western Central Asia in the beginning of the eighth century, and were fast becoming the neighbors of the Tibetans, who were expanding to the west. In 715 Tibet had concluded an alliance with the Arabs. Together with their new allies, Tibetan troops began to harass T'ang garrisons at the western end of the Silk Route at Kashgar. They also worked together to put a prince of their choice on the throne of Ferghana. According to an eleventh century Chinese history, they succeeded in 715. The relationship between Tibet and the Arabs must have been very friendly indeed, for historian al-Ya'qūbī reports that Tibet sent envoys to the Arab governor in Khurasan, the easternmost Arab province bordering on Central Asia, inviting masters of Islamic teachings to Tibet.

According to an eleventh century Chinese history, in 717 the Arabs and Tibetans joined together with the Turkish Türgish tribes led by Su-lo Khagan to attack Kashgar. Tun-huang records show that a Tibetan princess, Je-ba Dron-ma-lod, was sent in 734 to marry a Türgish khagan. This was apparently Su-lo Khagan (r. 717–737), for he was reported by the T'ang historians to have married a Tibetan princess.

BATTLES ON THE WESTERN FRONT

For Tibetan troops to reach Ferghana, over five hundred miles north of Ladakh, Tibet must have retained some hold on routes in the far west, despite losing the Silk Route. This hold was strengthened beginning about 721 when the Tibetan king received envoys from sTod-phyogs, the "upper regions" somewhere to the west beyond Zhang-zhung.

Tibetan military strategists had apparently developed a plan for expansion in spite of the T'ang presence in the west. They moved Tibetan troops along the Indus river, around the western end of the Silk Route, and into the far west. T'ang annals note that in 722 Tibet attacked Little Po-lu, the Gilgit region, which is known in Tibetan records as Bru-sha. T'ang records on the "Western Countries" indicate that Bru-sha bordered on the land of Oḍḍiyāna in the west and sBal-ti in the east. It was a critical region for the control of trade to the south, for through Bru-sha lay the most direct passage between the Central Asian oasis states, Kashmir, and India.

If Bru-sha were open to attack by Tibet, then Ladakh and sBal-ti (located between Ladakh and Bru-sha) must have already submitted to Tibet, for the road to Gilgit lay through these territories along the Indus river. Perhaps it was these areas that sent the sTod-phyogs envoys of 721. sBal-ti, the region around modern sKar-du, was known in the T'ang records of the "Western Countries" as Greater Po-lu. These records note that the country was just west of Tibet, touching Little Po-lu (Bru-sha), and that it belonged to Tibet. The Chinese pilgrim Hui-ch'ao, returning from India in 727, stated that Greater Po-lu, Yang-t'ung (Zhang-zhung?), and an unidentified region he called "So-po-tzu," were all controlled by Tibet.

The text about Khotan known as the 'Dri-ma-med-pa'i-'od-kyi-zhus-pa describes an invasion of the sBal-ti and Khotan region that seems to refer to this period (the date of the invasion is not given). Tibetan armies together with Sum-pa warriors defeated the sKar-du king named Īśvaravarman, as well as a Khotanese king. The sons of these two fallen kings then warred between themselves. Efforts were made to procure gold to ransom Khotan from the Tibetan armies, but Khotan was attacked, and numerous monasteries were damaged.

But the main focus of military attention at this time seems not to have been on Khotan so much as on Bru-sha. Because of its location, Bru-sha naturally became embroiled in a tug-of-war between the Tibetan armies and the T'ang armies. If the T'ang Empire held Bru-sha, it could block the Arab-Tibetan alliance and allow T'ang troops to link up with their ally in Kashmir, Lalitāditya, ruler of the Kārkoṭa dynasty. This monarch had sent to China in 733 for help against the Tibetans. Though China sent no aid, Lalitāditya managed to keep both the Tibetan and the Arab soldiers out of Kashmir, while also expanding the Kārkoṭa rule across northern India.

The open conflict between Tibet and T'ang China must have put the Chinese princess Kim-sheng in a difficult position. According to a Chinese encyclopedia published in 1013, she wrote secretly to the Kashmīri ruler in 724, asking for asylum, though apparently nothing came of the request.

In 736 Tibetan troops again attacked Bru-sha, and the Bru-sha ruler paid homage to Tibet, traveling to Brag-dmar to meet King Mes-ag-tshoms. That same year, T'ang forces attacked Bru-sha. But the ties with Tibet were strengthened in 740 when the Tibetan princess Khri-ma-lod married the king of Bru-sha. In 747 another large battle between Tibetan and T'ang troops took place. Under the Korean general Kao Hsien-chih, commander of the T'ang outposts in the West, T'ang troops fought off the Tibetans.

BATTLES ON THE EASTERN FRONT

Tibetan troops were also pressing on the eastern districts. In 720 they took 'Bug-cor, the Uighur principality at Sa-chou in the Tun-huang region. In 727 they attacked Kan-chou, an important outpost to the southeast of Tun-huang, and captured Kua-chou just east of Tun-huang, an important stop on the Silk Route. The Tun-huang chronicles explain that over many years, Kua-chou had become a repository of great riches. After Tibet raided this prosperous city, the simplest Tibetan folk were wearing Chinese silks. The following year, however, in extremely fierce fighting, T'ang troops drove off the Tibetan army after an eighty-day seige.

In 738 Tibet lost its hold on the Nine Loops of the Huang Ho, given as dowry. Two years later the great fort at An-jung north of Ch'eng-tu was also lost. A clever commander had established a secret pact with local people, who led the T'ang troops into the castle. For over sixty years, T'ang troops had tried to regain this borderpost. T'ang annals note that the emperor had been informed by military advisers that more than a million troops would be required to seize An-jung, and he himself participated in the secret plans. Another loss occurred in 749 when the famous T'ang general Ko-chu Han captured the Tibetan fort at Shih-p'u near the border in the Red Hills. This fort, known as mKhar-lcags-rtse, "Iron Peak Fort," was a formidable garrison perched high in the mountains like an eagle's nest.

Despite these losses, the Tibetan military maintained their long-range strategy to recapture the Silk Route. The plan seems to have been to envelop the Tarim basin by expeditions in both the east and the west at the same time. Their armies must have been extremely large and well organized to attempt such a feat, which called for year after year of warfare on two distant fronts. It was to be forty years before the whole plan succeeded, but in the meantime, Tibetan territory and power kept growing.

TIBETAN INFLUENCE EXPANDS

After many years of struggle and bloodshed in the far northwest, Mes-ag-tshoms again sent armies north of the Jaxartes river. There they helped the Arabs and the Qarluq Turks defeat the T'ang army at the battle of Talas in 751. According to Western historians, this crucial battle ended T'ang influence in the west.

According to Tun-huang and T'ang records, Tibetan influence increased. In 756, envoys from sTod-phyogs, the "upper regions" in the west, again came to Tibet from the Black Ban-'jag, Gog, and Shig-nig. Gog seems to be somewhere near Bru-sha, and Shig-nig is identified by F.W. Thomas as Shignān, a principality in the Pamirs. Ban-'jag, according to Uray, might be the Bunji area near Gilgit.

Toward the end of Mes-ag-tshoms' reign, Tibetan influence greatly expanded to the southeast into 'Jang (Nan-chao) kingdom, which was centered around Ta-li lake in modern-day Yunnan. In the iron-hare year of 751, the Tibetan king received the ruler of Nan-chao, Kag-la-bon (r. 748–779), known in the Chinese annals as Ko-lo-feng. He was the son of Pi-lo-ko (r. 729–748), who had united many of the Nan-chao tribes. These people were "Tibeto-Burmese," according to some scholars, though others believe they were Thai or were ruled by a Thai upper class.

After meeting Mes-ag-tshoms, Kag-la-bon became his close ally, formally offering his country to the Tibetan king. In 753 Mes-ag-tshoms sent Kag-la-bon a golden diplomatic seal. In 754 and 756 the king dispatched Tibetan troops to aid Nan-chao against T'ang armies, according to the Tun-huang documents and Chinese sources on Yunnan history.

During a celebration of the new relations with 'Jang, poetry and songs were composed. Recorded in the Tun-huang chronicles, they proclaim the Tibetan king to be unlike any other king, a son of the gods, who could unite all the princes and protect the people. The 'Jang commanders are hailed as unusually brave warriors, their princes as heroic, and their decision to join with Tibet as wise indeed.

Nan-chao must have been a useful ally to Tibet, for this kingdom soon dominated not only eastern Yunnan and western Kweichow, but also the upper valley of the Irrawaddy river and parts of northern Burma and Assam. This strategic position gave Nan-chao control of the trade routes between China and Tonkin to Burma and India. In 779 I-meu-sin succeeded Kag-la-bon, and maintained the alliance with Tibet until 794.

CULTURAL AND RELIGIOUS DEVELOPMENTS

Although his reign was filled with the complexities of politics and foreign relations, Mes-ag-tshoms found time for the advancement of Tibetan scientific and religious thought. According to the Shes-bya-kun-khyab and the mKhas-pa'i-dga'-ston, the king supported the translation of medical texts as had his predecessor Srong-btsan-sgam-po. He invited to Tibet a physician in the Greek tradition, who became the court doctor. This famous physician, Biji Tsampaśilaha, translated not only Western medical texts, but also texts from China and India, and is sometimes referred to as the "Chinese scholar." He was also responsible for promulgating a physician's oath similar to the Hippocratic oath of the Greek tradition. Biji remained court physician under the following king, Khri-srong-lde-btsan, and three of his students established teaching lineages that continued into modern times.

Mes-ag-tshoms also made efforts to bring more Buddhist teachings to his people. He had discovered the prophecies written on a copper plate by Srong-btsan-sgam-po, and thought he might be the future king who could help spread the Dharma in Tibet. His Chinese queen, a devout Buddhist, must also have been an inspiration to the king. Soon after her arrival in Tibet, Kim-sheng had located the Jo-bo-chen-po brought by 'Un-shing Kong-jo. The statue had been concealed in the Ra-sa-'phrul-snang temple, but the queen had it brought forth. The Ngo-mtshar-rgya-mtsho notes that she also instructed the people in special ceremonies for the sick and dying, and promoted Buddhist practices.

Interested in obtaining more Buddhist texts, Mes-ag-tshoms had heard that two great Buddhist masters, Buddhaguhya and Buddhaśānti, were residing in mNga'-ris. Messengers were sent to their hermitage at Mount Kailāśa (Ti-se), extending them an invitation to visit lHa-sa. The masters did not visit the king, but gave sutra teachings together with kriyāyoga and upāyoga teachings to the messengers. They returned to the king with these teachings, which he had preserved in specially designed books. To properly house these texts, the king had five temples built at the following locations: Brag-dmar-mgrin-bzang, Brag-dmar-ke-ru, gSang-mkhar-grags, mChims-phu-sna-rla, and a fifth temple within his own palace.

Mes-ag-tshoms also dispatched Sang-shi to China to obtain texts. According to the sBa-bzhed, the Chinese emperor had sent Sang-shi to be a companion to his grandson, the young prince Khri-srong-lde-btsan. When Sang-shi returned to China, the emperor wanted him to remain as a minister of state, but he instead asked for the religious texts desired by his Tibetan king. Sang-shi also traveled to the holy mountain of Wu-t'ai-shan in northern China where he received texts as well. But he did not return to Tibet until after Mes-ag-tshoms had died.

During the reign of this king, various works on mathematics and calculation were translated, as well as Buddhist texts. Mūlakośa of Blan-ka and gNyags Jñānakumāra translated the Karmaśataka-sūtra and the Suvarṇaprabhāsottama-sūtra.

According to the Blue Annals, monks were also invited from Khotan and treated with respect, but no ordinations took place. The Li'i-yul-lung-bstan-pa confirms this, saying that in the time of the Chinese princess Kim-sheng, the Li-yul king banished Buddhist monks, who then traveled via Lop Nor into Tibet. The Tibetan king built seven monasteries for monks arriving from Ferghana, Tokharia, 'An-se (Bukhara), Shu-lig (Kashgar), Kashmir, and Gus-tig near Samarkand. As the Muslim Arabs moved into these regions, some Buddhist monks fled to near-by Bru-sha, which was under Tibetan control. When the king heard of this situation, he invited these monks to Tibet as well.

About the same time that Mes-ag-tshoms was building temples at Brag-dmar and various other locations, the Kashmīri king Lalitāditya was founding Rājavihāra. It would be interesting to know more about Mes-ag-tshoms' relationship with Lalitāditya, for though they were political rivals whose territorial claims were in conflict, they both supported the Buddhist teachings. Future research might also reveal contacts between Tibet and eighth century Buddhists in Bengal, Nan-chao, and Burma. The trade routes between these lands, all of which were Buddhist, were in the hands of Tibet's ally Nan-chao.

Sometime between 739 and 741, the Princess Kim-sheng fell ill with smallpox and soon passed away. The pestilence spread rapidly, and a number of children died. The ministers of state interpreted this as a consequence of the king and queen's support for Buddhism. The Li'i-yul-lung-bstan-pa notes that the ministers begged the king to expel the Buddhists, and the king begged the ministers to consider whether such an act was necessary. Finally the monks were asked to leave and traveled to Gandhāra and Kośāmbī in India.

How the ministers finally overrode the king's desire to establish a Buddhist Sangha is not explained. Serious, reserved, and somewhat ascetic, Mes-ag-tshoms was often criticized for not participating in hunting expeditions and festivities. Some of his ministers despaired of the king's demeanor and appearance, which came increasingly to resemble that of an Indian holy man. The king may have felt it unwise to persist openly in trying to change the old ways of his people after the popular Buddhist queen had passed away.

WIVES AND HEIRS OF MES-AG-TSHOMS

Besides his Chinese wife, Mes-ag-tshoms married a noblewoman from sNa-nam, Mang-mo-rje bZhi-steng. Another wife was lCam lHa-spangs, who passed away in 730, according to the Tun-huang annals. Mes-ag-tshoms married the noblewoman Khri-btsun from 'Jang, perhaps in order to cement the alliance with Nan-chao, but we could find no record of the date of this marriage. She may actually have been from the Mya-ba tribes who had submitted to Tibet in 703. A son, lJang-tsha lHa-dbon, was the child of Queen Khri-btsun, according to the sBa-bzhed. This account claims that lHa-dbon was the husband intended for the Chinese princess (see above). The Tun-huang annals note that a son lHas-bon died in 739. Whether this is lHa-dbon is not clear.

The king's second son, Khri-srong-lde-btsan, was born at Brag-dmar-mgrin-bzang north of Yar-lung, not far from the site where the famous monastery of bSam-yas would soon be founded. His birth took place in the water-horse year of 742, according to the Tun-huang documents, but many later histories give an iron-horse year of 730. The commonly accepted tradition, found for example in the sBa-bzhed and Bu-ston, states that the Chinese wife was his mother, and that the sNa-nam queen tried to deceive the court, claiming for years that the child belonged to her. When the child was asked to declare which one was his mother, he unhesitatingly went to the side of the Chinese princess.

Tun-huang documents, however, maintain that Khri-srong-lde-btsan's mother was the sNa-nam wife, and that the Chinese princess died in 739. The T'ang annals record the death of the Chinese princess in 741, which is still before Khri-srong-lde-btsan's birth in 742, unless the earlier date of 730 is accepted. The disagreement among these records may be connected to the deception perpetrated by the sNa-nam queen.

Mes-ag-tshoms was murdered in 755 during a palace revolt. The king's interest in the Dharma certainly provoked the wrath of his ministers, but whether there were other areas of dispute is not known. The death of young Khri-srong-lde-btsan was also part of the plot. Even as a child, Khri-srong-lde-btsan had a remarkable firmness of mind that would have made it very difficult for the ministers to control him completely during his minority. They knew the young prince was already interested in Buddhist teachings, for Mes-ag-tshoms had discussed this with them, trying to convince his cabinet that the Dharma's success was inevitable.

The prince, however, escaped harm, thanks to a faithful minister. An edict later erected at Zhol in lHa-sa, which can still be seen today, states that the ministers 'Bal lDong-tsab and Lang Myes-zigs were responsible for the death of the king. The life of Prince Khri-srong-lde-btsan was saved by Klu-khong, and the inscription records his deeds and the granting of privileges to his family and descendants.

In his fifty-one-year rule, Mes-ag-tshoms expanded the empire into the key regions of Central Asia and strengthened Tibetan foreign relations. He brought cultural influences of many kinds into Tibet, and established an alliance with the Arabs, another marriage alliance with T'ang China in the east, and an alliance with 'Jang (Nan-chao) in the south. By the middle of the eighth century, Tibet had become one of the great powers in Asia.

Songs in the Tun-huang chronicles show how the Tibetans of the eighth and ninth centuries felt about their land and their kings:

> "Coming from the land of the gods,
> from the seven stages of the blue heavens,
> the sons of the gods protect the people.
> Among all the kingdoms of men,
> never has there been one like ours!"

ARAB EXPANSION

Inspired by the teachings of Mohammed, the Arab tribes united in the early seventh century, and soon expanded far beyond the Arabian peninsula. By 750 A.D. the Arab Caliphate controlled the old territories of Persia and part of the Roman Empire (Byzantium). North Africa fell to the Arabs as well, and Europe was threatened. Spain was taken in 712, at the same time that thrusts were being made into Central Asia where the Arabs were now confronted with a new challenge: the armies of the Tibetans, the Turks, and the Chinese.

TRIPLE ALLIANCE

In the eighth century Tibet formed a powerful triple alliance in the far northwest where Tibet's empire met neighboring forces of the Turks to the north and the Arabs to the west. Arab forces had reached the Oxus river by 683 and Ferghana by 705. Tibetan and Turkish troops were in Ferghana in 700. In 715 Arab and Tibetan armies joined together to attack Kashgar, one of the most important garrisons on the Silk Route, threatening T'ang hold on Central Asia.

THE LAND OF BRU-SHA

The kingdom of Bru-sha northwest of Tibet was a critical region for control of Central Asian trade with India and Kashmir. Located between the Arabs to the west, the Turks to the north, China to the east, Kashmir and Tibet to the south, Bru-sha was caught in a long struggle between Tibetan and T'ang troops for twenty years in the eighth century. If Bru-sha were lost, a T'ang commander declared, all of the Western Regions would fall into the hands of Tibet.

T'ANG POWER DECLINES

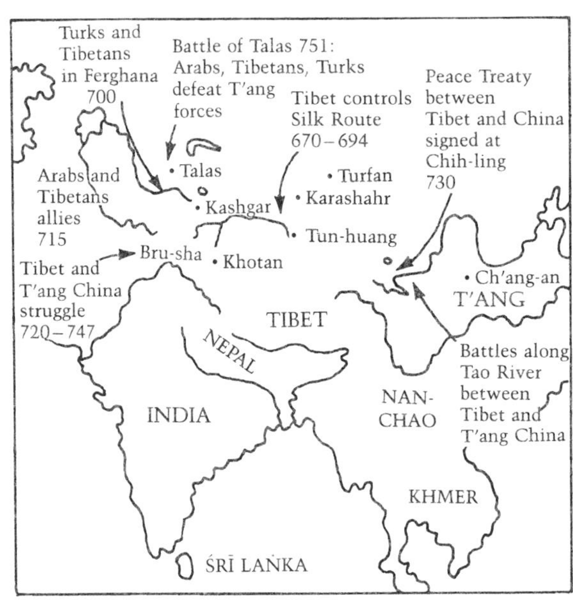

Throughout the seventh century T'ang power across Central Asia had steadily grown. But the Tibetan army was now gradually taking control of Central Asia. The T'ang Empire held the Silk Route through Khotan and Kucha since taking it from Tibet in 692. The battle of Talas ended T'ang power in the west. Bru-sha and sBal-ti became Tibetan territories; over the next forty years, the long-term Tibetan strategy to regain the Silk Route was to succeed.

T'ANG EMPERORS CONTEMPORARY WITH DHARMA KINGS

618–626 A.D.
Kao-tsu

627–649 A.D.
T'ai-tsung

649–683 A.D.
Kao-tsung

684–705 A.D.
Empress Wu

705–710 A.D.
Chung-tsung

710–712 A.D.
Jui-tsung

712–756 A.D.
Hsüan-tsung

756–762 A.D.
Su-tsung

762–779 A.D.
Tai-tsung

780–805 A.D.
Te-tsung

805–820 A.D.
Hsien-tsung

820–824 A.D.
Mu-tsung

824–827 A.D.
Ching-tsung

827–840 A.D.
Wen-tsung

840–846 A.D.
Wu-tsung

846–859 A.D.
Hsüan-tsung

THE ERA OF MES-AG-TSHOMS

TIBET	ASIA	FAR EAST	EUROPE
704–755 A.D. Mes-ag-tshoms rules Tibet	705 A.D. Beginning of Arab conquest of western Central Asia under General Qutaiba	710–784 A.D. Nara period in Japan; Buddhist temples founded	711 A.D. Muslims invade Spain; Golden Age of Spanish Jewry begins
710 A.D. Marriage alliance between Tibet and China	712 A.D. Arabs enter Sindh in N.W. India	712–756 A.D. Hsüan-tsung rules T'ang China	717–741 A.D. Emperor Leo III of Byzantium reorganizes empire; opposes use of images in Christian churches
715 A.D. Arab-Tibetan alliance	724–760 A.D. Lalitāditya extends Kārkoṭa rule to Bengal; prevents Tibetans and Arabs from taking Kashmir	Golden Age of Chinese poetry and painting	718 A.D. Constantinople defends against Arabs
720–740 A.D. Tibet gains control of both Bru-sha and sBal-ti in far west	730 A.D. Yaśovarman rules Kanauj	724–749 A.D. Emperor Shōmu rules Japan	730 A.D. Pope Gregory II excommunicates Byzantine Emperor Leo III
727 A.D. Tibetan troops attack Kan-chou and Kua-chou	740 A.D. Gurjara-Pratihāra dynasty defends India against Arab incursions	725 A.D. Academy of Letters established in China	732 A.D. Arab expansion in Europe halted by Frankish king Charles Martel
730 A.D. Treaty between Tibet and China	740 A.D. Shi'ite revolt in Persia	732 A.D. Manicheanism condemned in China	739 A.D. Pope Gregory III asks Martel's aid against Arabs
734 A.D. Marriage alliance with Turks	745 A.D. Uighur Empire established at Orkhon, Mongolia; N. and W. Turkish lands seized	747 A.D. General Kao leads T'ang army into western Central Asia	742 A.D. Birth of Charlemagne
742 A.D. Birth of Khri-srong-lde-btsan	747 A.D. Abu Muslim revolts against Umayyads; establishes Abbasid Caliphate	751 A.D. Battle of Talas: end of Chinese power in Central Asia	755 A.D. Byzantium at war with Bulgars
747 A.D. T'ang forces drive Tibetan troops from Bru-sha		755 A.D. Revolt of An Lu-shan, frontier general, begins civil war in China; T'ang troops withdrawn from Central Asia	774 A.D. Frankish kingdom includes modern lands of France, Germany, N. Italy
751 A.D. Nan-chao ('Jang) submits to Tibet			
751 A.D. Battle of Talas: Tibetan, Arab, and Turk armies defeat Chinese			

CHAPTER SEVENTEEN

THE DHARMA IS PROCLAIMED

In the middle of the eighth century, the land of Tibet finally opened wide its doors to the Dharma due to the efforts of Khri-srong-lde-btsan. He is revered by the Buddhist tradition as a great Dharma King and an incarnation of Mañjuśrī. Even in his own lifetime, he was recognized as a remarkable individual by outstanding teachers such as Padmasambhava and Śāntarakṣita. The Buddhist Master Buddhaguhya wrote King Khri-srong-lde-btsan a letter that addressed him as follows: "Homage to Mañjuśrī, the king of the black-headed Tibetans, the son of Ag-tshom Mes, grandson of 'Dus-srong, who continues without interruption the lineage of Bodhisattvas that began with King Srong-btsan-sgam-po, the incarnation of Avalokiteśvara."

HOSTILITY TOWARD THE DHARMA

In the early part of his rule, which began in 756 when he was thirteen, Khri-srong-lde-btsan could accomplish little for the Dharma, for he was confronted with powerful anti-Buddhist ministers. In an edict written by the king himself, preserved in the mKhas-pa'i-dga'-ston, Khri-srong-lde-btsan noted that these ministers had responded to Mes-ag-tshoms' support of the

Dharma by having a law written forbidding the practice of Buddhism. Intent on destroying the influence of the Dharma, they banished two ministers who were pro-Buddhist, and they persecuted Dharma practitioners.

The hostile ministers attempted to move the Jo-bo-chen-po, the Buddha statue brought from China by Srong-btsan-sgam-po's Chinese queen. But the statue could not be lifted! The ministers then turned the gTsug-lag-khang temple into a slaughter-house, and hid the statue beneath sand. After many misfortunes befell those who had buried the Jo-bo, it was uncovered and taken out of lHa-sa to Mang-yul. When the messenger Sang-shi finally returned from China with the texts Khri-srong-lde-btsan's father had requested, he hid them for safekeeping, for he dared not present them to the young king.

In the face of such opposition, young Khri-srong-lde-btsan realized he needed a fuller understanding of the political and religious conditions in his country. He read the biographies of his father and his grandfather with great care, studying the events of the past. Finding himself surrounded by relatives and powerful ministers whose devotion to the old traditions of Tibet were very strong, the king recognized the depth of these roots and how carefully he must proceed if he wished to open up a new vision for his people. In fact some seven years were to pass before he would be able to invite Buddhist teachers to Tibet. Skillfully the king made his way around obstacle after obstacle.

The young king first gave the position of minister in Mang-yul to sBa gSal-snang, a man with great faith in the Buddhist tradition, who wished to travel to India in search of teachings. He also had the texts from China quietly brought forth from hiding, and the Chinese Mes mGo, together with the Paṇḍita Ananta and Sang-shi, set about translating them. Reading the texts, even without the instruction of a teacher, the king was filled with faith, for his natural intelligence recognized truth, and his understanding was deep.

When obstructions and threats from the anti-Buddhist ministers continued, Khri-srong-lde-btsan temporarily discontinued translations. But he began to make preparations to invite a Buddhist master to Tibet. After sending Sang-shi to Mang-yul near Nepal for safety, he dispatched sBa gSal-snang to Nepal and India. After first visiting Mahābodhi and Nālandā, gSal-snang traveled to Nepal to invite Śāntarakṣita, the famous abbot of the Buddhist university of Vikramaśīla, to teach in Tibet.

ŚĀNTARAKṢITA COMES TO TIBET

Returning to Tibet, gSal-snang, now known as Jñānendra (the religious name given him by Śāntarakṣita), spoke to the king of the great master, and the king longed to meet him. But first the anti-Buddhist ministers had to be removed from power. The king sent gSal-snang to safety in the countryside, while a plan was devised by the pro-Buddhist ministers. They bribed an oracle to predict misfortune for the kingdom if the two greatest among the ministers did not ward off this danger by undertaking a special retreat in the royal tombs. The chief anti-Buddhist minister had long been hailed as the most outstanding by the faction opposed to the Dharma. When one of the pro-Buddhist ministers leaped up to offer himself as ransom, the hostile minister had to join him. The two were sealed inside the tombs, but the king's supporter had prearranged his own escape.

The king then sent several trusted ministers to speak directly with Śāntarakṣita; Khri-srong-lde-btsan wanted to be certain the teachings were genuine and that this master taught in a way his people could understand. Receiving good reports from his envoys, the king renewed his invitation to the renowned abbot, and journeyed to meet him at Brag-dmar 'Um-bu-tshal. In the presence of this great Buddhist master, Khri-srong-lde-btsan recalled their past lives together, and his intention to establish the Dharma in Tibet grew even stronger.

Śāntarakṣita taught for four months in the king's palace, working through the Kashmīri interpreter Ananta. The abbot explained basic Buddhist philosophy and ethics to a small group assembled by the king, and Ananta translated texts.

Some of the ministers, however, became very uneasy at Śāntarakṣita's presence. A series of disasters were interpreted as great displeasure on the part of the old Tibetan gods. Knowing that his ministers' minds were completely convinced by these events, the king reluctantly agreed to ask Śāntarakṣita to return to Nepal. But he also sent sBa gSal-snang and Sang-shi to China, where they met with the emperor and studied with Buddhist teachers. Receiving a prediction from a Chinese monk that they would be able to help establish the Dharma in Tibet, they returned to Tibet with many texts, which they buried for safekeeping.

PADMASAMBHAVA ARRIVES IN TIBET

Though repeatedly opposed by some of his ministers, King Khri-srong-lde-btsan continued his efforts to establish the Dharma successfully in his land. But he knew that the sacred status of a Tibetan king depended on an ancient tradition that could not be abruptly altered without the king losing the very basis of his power. Endangering his own position would also throw the country into turmoil and destroy any opportunity for bringing the Dharma into Tibet.

To accomplish the difficult task of changing the outlook of his people, the king planned to invite Padmasambhava to Tibet. This great Tantric master had been recommended by Śāntarakṣita for his ability to subdue negative forces. Padmasambhava was renowned as the Lotus-born Guru from Oḍḍiyāna, a very ancient and sacred place connected with the Vajrayāna teachings for many centuries. Located by most modern historians in the Swat valley southwest of Bru-sha (Gilgit), Oḍḍiyāna had been visited in the seventh century by the Chinese pilgrim Hsüan-tsang, who had seen over a thousand old temples in the hills.

Khri-srong-lde-btsan now sent five men to invite Guru Padmasambhava. The Guru knew the king was in need of him, and the messengers found him already in Mang-yul near the Nepal border. As he proceeded toward the capital, negative forces rose up to obstruct his entry into Tibet, but the great Guru converted them all. When Padmasambhava and Khri-srong-lde-btsan met, the king became his disciple. Khri-srong-lde-btsan's second edict notes that it was in his twentieth year that he received the Buddhist teachings, and later sources agree that he was twenty when he met Padmasambhava and Śāntarakṣita. This would be in 762 if we accept 742 as the king's birthdate.

Once Padmasambhava was at work in Tibet, Śāntarakṣita also returned. The two great teachers, together with the king, then began building the temple of bSam-yas, which Śāntarakṣita had attempted to build on his previous visit. Every effort made during the day had at that time been destroyed during the night by demons. This time, Padmasambhava called forth the demons and subdued them, converting their energy to assist rather than oppose the Dharma.

DATING THE FOUNDATION OF BSAM-YAS

Various dates for the founding of bSam-yas are given in the Buddhist tradition, partly because the events marking the founding and completion are understood differently. Bu-ston states the work began in the fire-hare year of 787 and was finished in an earth-hare year twelve years later in 799. The Deb-ther-dmar-po also gives an earth-hare year as the completion date. But a date of 799 does not fit with the dates of the famous debate between the Indian and Chinese schools of Buddhism that was held after bSam-yas was completed. According to Demiéville's research, the debate took place between 792 and 794.

Other sources also note the temple was begun and completed in hare years, but they do not specify which ones. If Khri-srong-lde-btsan was born in 742, likely hare years for the founding might be 763 when the king was twenty-one, or 775 when he was thirty-three. A number of histories state that the temple was begun when the king was twenty, soon after Padmasambhava's arrival, so a period of construction from the water-hare year 763 until the wood-hare year of 775 would seem to be acceptable.

But there is another tradition that states bSam-yas was founded in a tiger year, completed in five years in a horse year, and consecrated in a sheep year. One of Khri-srong-lde-btsan's edicts also states that bSam-yas was consecrated in a sheep year. The Blue Annals gives hare year to sheep year as starting and completion dates, which might fit this tradition if the consecration were considered the time of actual completion. Likely dates might then be from the water-tiger year of 762 until the fire-horse year of 766, with the consecration taking place in 767, a fire-sheep year.

Additional support for a founding date of 762 or 763 appears in the Tun-huang chronicles. Although no dates are given, these records mention the establishment of Buddhism and the building of temples just before referring to the capture of the T'ang capital, which can be dated with confidence to 763 from both T'ang and Tibetan records.

BUDDHISM BECOMES THE RELIGION OF TIBET

Whether completely finished in twelve or five years, the bSam-yas monastery was a magnificent complex, constructed in mandala form and modeled on Odantapurī in India. It was located south of lHa-sa and north of the gTsang-po and the Yar-lung valley. The main temple represented Mount Sumeru, the central axis of the world in Indian cosmology. It was surrounded by four major and eight minor temples, symbolizing the continents, and two additional temples representing the sun and the moon. A large wall topped with stupas ran around the perimeter of the temple grounds with gates in the four directions marked by four large stupas. Several queens of Khri-srong-lde-btsan assisted by sponsoring the building of some of the temples. dMar-rgyan built the Khams-gsum-bzang-khang-gling, and Pho-yong-bza' the dBus-tshal-gser-khang-gling.

The founding and completion of bSam-yas inspired the king with great confidence that Tibet could indeed become a Buddhist land, despite the initial difficulties he had faced. Inscribed on a bell at bSam-yas is a declaration of the king's support for the Buddhist teachings; a short edict by Khri-srong-lde-btsan carved on a pillar at bSam-yas declared Buddhism as the official religion of Tibet.

Two longer edicts are preserved in the mKhas-pa'i-dga'-ston. One was signed by the ministers, relations of the king, generals, and various nobles, who swore to uphold the Dharma. These oaths were originally written in gold on blue paper and kept at bSam-yas, while thirteen copies were distributed to various monasteries, as far away as Bru-sha, Zhang-zhung, and mDo-smad. This edict must have been proclaimed sometime between 768 and 782 because the chief minister signing the oath was Zhang rGyal-gzigs. rGyal-gzigs, one of the generals who captured Ch'ang-an (see chapter 18), held the post of chief minister beginning sometime after 768 until 782.

Once bSam-yas had been finished, twelve Sarvāstivādin monks from Kashmir were invited to Tibet, and seven young Tibetans were selected to become the first monks. Known as the sad-mi-bdun, the seven "test cases," these men were greatly successful as monks, demonstrating that Tibetans could indeed practice the rigorous Buddhist discipline. Ordained by Abbot Śāntarakṣita, they became the holders of the first Vinaya lineage in Tibet. Three hundred people, together with two of Khri-srong-lde-btsan's queens, decided to become monks and nuns, and the king established special laws that provided for their support.

With the establishment of this foundation for Buddhist practice, Tibetan translators began working with masters invited from abroad, including Śāntigarbha, Viśuddhasiṁha, and the Kashmīri masters Jinamitra and Dānaśīla. A large number of texts were translated, including Vinaya, Sutras of all Three Turnings, and Tantras, as well as śāstras on all subjects: Prajñāpāramitā, Cittamātra, Madhyamaka, Tantras and Sutras, Vinaya, Logic, Abhidharma, and Grammar. In 781 Khri-srong-lde-btsan also invited monks from China to visit Tibet on a regular basis.

Bu-ston explains that different kinds of Dharma work were done in different locations at bSam-yas, which was a huge complex of numerous temples, libraries, and meditation halls. Monastic discipline was taught in the rNam-dag-khrims-khang-gling; the Chinese master taught meditation in the Mi-g·yo-bsam-gtan-gling; grammars and dictionaries were written in the brDa-sbyor-tshangs-pa'i-gling; and treasures were stored in the dKor-mdzod Pe-har-gling. Copies of the translated texts were preserved in several locations. A catalogue prepared in a dragon year at the palace of lDan-dkar listed titles of the texts that had been translated, including the number of chapters and verses contained in each text. This catalogue is preserved in the bsTan-'gyur.

As the teachings began to spread throughout Tibet, the king built twelve special meditation centers and established meditative retreat centers at Yer-pa and mChims-phu. He offered his support to these practitioners, who became famous as accomplished siddhas. The king himself wrote the bKa'-yangs-dag-ma-tshad-ma, an extraordinary exposition of the Dharma, which clarified difficult points and conveyed his deep personal understanding of the Buddhist teachings.

Though the court chronicles found at Tun-huang rarely mention Buddhism, they do briefly describe the establishment of the Dharma:

> "The incomparable religion of the Buddha was received,
> and monasteries were built everywhere,
> in the central regions and even in the frontiers.
> Once the doctrine was established,
> people were filled with compassion
> and released from birth and death."

VAIROTSANA AND VIMALAMITRA

Khri-srong-lde-btsan now sent Vairotsana, the Tibetan translator, to India where he studied with prominent masters and collected a large number of teachings and texts. Vairotsana also traveled to China and other northern regions such as Khotan, where he was invited by the Queen Byang-chub-grol-ma. The

Li'i-yul-lung-bstan-pa reports that a monk known as Vairotsana arrived in the reign of the fifty-fifth king, Vijaya Sambhava, who apparently ruled in the eighth or ninth century.

Returning to Tibet, Vairotsana privately taught Khri-srong-lde-btsan and a few others chosen by the king. Disputes arose, however, about the validity of his teachings, and skeptics spread rumors about Vairotsana and the queen rMa-rgyal. The king saw the necessity of exiling Vairotsana to the eastern province of Khams to avoid further difficulties. There Vairotsana stayed in rGyal-mo-rong with a local ruler known as mDo-bzher-nag-po. He continued to teach in the east and attracted several excellent disciples, including gYu-sgra-snying-po.

Khri-srong-lde-btsan then invited to his court the outstanding Oḍḍiyāna master Vimalamitra, who had studied with the greatest Vajrayāna masters of the time. Vimalamitra taught and translated at bSam-yas, working closely with the disciples of Padmasambhava. When he reassured the king that the teachings given by Vairotsana were authentic and very valuable, Vairotsana was asked to return. He became known as the king of translators and the most learned among the Tibetans. Vairotsana and the master Vimalamitra were assisted in translating by gNyags Jñānakumāra, sKa-ba dPal-brtsegs, Cog-ro Klu'i-rgyal-mtshan, gYu-sgra-snying-po, and rMa Rin-chen-mchog.

Over the next years, the Vajrayāna teachings brought to Tibet by Padmasambhava, Vimalamitra, and Vairotsana were widely spread. Ye-shes-mtsho-rgyal, gNyags Jñānakumāra, and gNubs Sangs-rgyas-ye-shes were especially important in establishing the Inner Tantra lineages in Tibet. These teachings, which were continuously transmitted, later became known as bka'-ma.

Other valuable teachings, ritual objects, and precious items were carefully hidden by Padmasambhava and his close disciple mTsho-rgyal. These sacred treasures, known as gter-ma, would be discovered in later times by reincarnations of Padmasambhava's disciples. With his blessing, they would be able to locate and translate the gter-ma that preserved the direct teachings of Padmasambhava for future generations.

KHRI-SRONG-LDE-BTSAN: DHARMARĀJA

The reign of Khri-srong-lde-btsan marked the beginning of the formal acceptance of the Buddhist teachings as the religion of the land. Bon histories often explain the difficulties faced by the Dharma Kings as a result of abandoning the old Bon practices. But Khri-srong-lde-btsan interpreted events in a different light, for he had total confidence in the teachings of the Buddha.

One of his edicts expressed his certainty that following the Dharma would bring immeasurable blessings to Tibet. There he tried to explain to the people the importance of understanding karma, of knowing what actions are truly beneficial for oneself and others. The edict notes that the king met with important local chiefs and discussed the Dharma with them, pointing out three good reasons for practicing Buddhist teachings. One could read and study the teachings and see for oneself how profound they were. Or one might be impressed with the example of the previous Tibetan kings, who had for generations supported the Buddhist teachings. Or one might look to the visiting Buddhist masters—do they not inspire confidence? The king summarized his complete confidence in the teachings of the Buddha in this way: "From the Dharma, nothing but good can possibly arise."

The Bon tradition sometimes refers to a persecution during the reign of this king. Khri-srong-lde-btsan did limit certain activities of Bon teachers and even banished some of them. But he was very careful to protect valuable Bon teachings, preserving them as treasures. Khri-srong-lde-btsan's respect for the ancient traditions of his country is revealed in an inscription at 'Phyongs-rgyas, which states that his rule was in accord with the customs of his ancestors. Like the kings before him, he is called 'Phrul-gyi-lha, lHa-sras, and lHa bTsan-po, the son of the gods, the magical king. But he was given one additional epithet—Byang-chub-chen-po, Great Enlightened One.

After the master Śāntarakṣita passed away, a great debate was held at bSam-yas to clarify the doctrine. Formal debate had been used in India for centuries to test opposing views, with the loser often converting to the view of the winner. In Tibet the teachings propagated by the Chinese Buddhists seemed to differ from those taught by the Indian masters. Thus, in 792 Khri-srong-lde-btsan invited the Indian master Kamalaśīla, a disciple of Śāntarakṣita, to defend the Indian view and a Chinese monk known as Hwa-shang Mahāyāna, to put forth the Chinese view. When the Indian system proved superior, the king proclaimed it the official doctrine, and the Chinese teachers were asked to leave Tibet. After the debate the king announced great endowments for bSam-yas to provide for the monks and abbot.

Khri-srong-lde-btsan was a true Dharmarāja, a ruler inspired by the teachings of the Dharma to bring benefit to his people and to his land. His deep desire to obtain for Tibet the best teachings available in all of Asia led to the founding of bSam-yas and his invitation to the finest Buddhist masters of the eighth century. Building carefully upon the foundation laid a century and a half earlier by Srong-btsan-sgam-po, Khri-srong-lde-btsan can be said to have established the teachings that nourished Tibetan civilization for over twelve hundred years.

Chos-rgyal Khri-srong-lde-btsan

BUDDHIST MASTERS INVITED TO TIBET

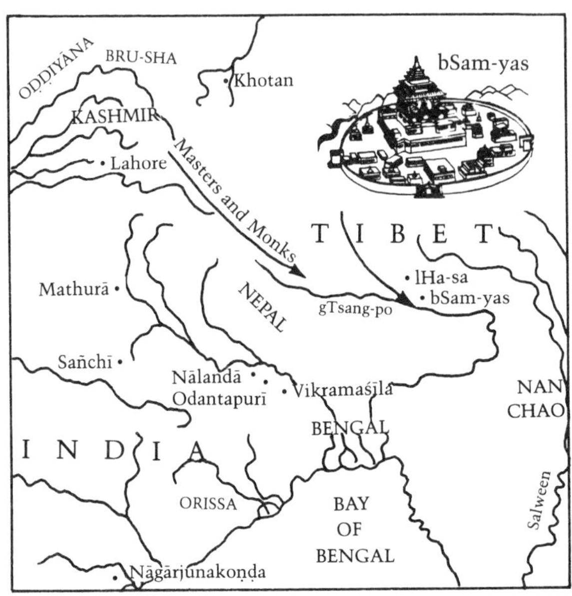

Intent on establishing the Dharma in his own country, the king Khri-srong-lde-btsan invited Abbot Śāntarakṣita from Vikramaśīla, a famous university in India; he also invited Padmasambhava, the Tantric master from Oḍḍiyāna. After bSam-yas was completed, teachers were invited from Kashmir and the Vajrayāna master Vimalamitra also arrived. Tibetan monks were now ordained, and translations of Buddhist texts were rapidly accomplished with the guidance of masters.

EIGHTH CENTURY TEMPLES

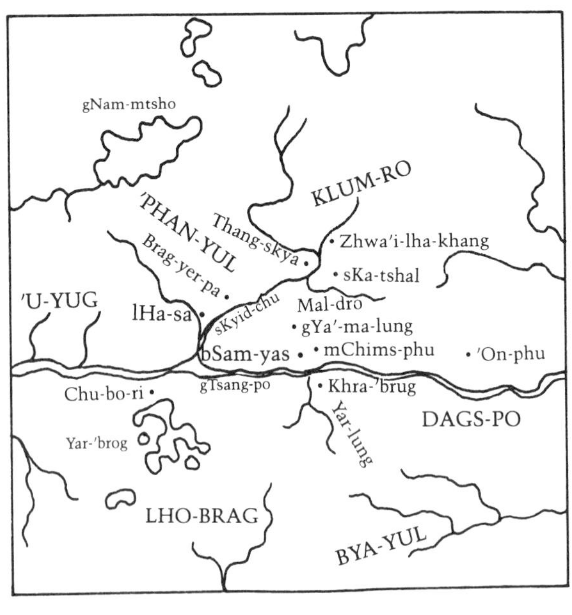

During the reign of King Khri-srong-lde-btsan, the first Buddhist monastery was founded at bSam-yas south of lHa-sa and north of Yar-lung by the king together with the Oḍḍiyāna Guru Padmasambhava and the Abbot Śāntarakṣita. Centers were built at Brag-yer-pa, mChims-phu, and other sites. The building of bSam-yas established the teachings of the Buddha in Tibet. The king proclaimed Buddhism the religion of the land, and Tibetans were ordained as monks.

BUDDHISM IN ASIA

INDIA AND KASHMIR	DEVELOPMENTS IN TIBET	CHINA CENTRAL ASIA	JAPAN, JAVA KOREA
Mahācārya Śāntarakṣita Abbot of Vikramaśīla Oḍḍiyāna Guru Padmasambhava Mahāpaṇḍita Vimalamitra Mahāpaṇḍita Jinamitra Mahāpaṇḍita Dānaśīla Mahāpaṇḍita Surendrabodhi Mahāpaṇḍita Haribhadra, disciple of Śāntarakṣita Mahācārya Śāntideva Mahāpaṇḍita Vinītadeva Mahāpaṇḍita Jñānagarbha, teacher of Śāntarakṣita Mahāvidyādhara Śākyamitra Mahāvidyādhara Līlavajra Mahāvidyādhara Buddhaguhya Vidyādhara Buddhajñānapāda Mahāsiddha Kṛṣṇa-pa Mahāsiddha Virūpa	762 A.D. Khri-srong-lde-btsan invites Śāntarakṣita and Padmasambhava 762/763 A.D. Founding of bSam-yas monastery 767/775 A.D. Completion of bSam-yas Edict of Khri-srong-lde-btsan proclaims Buddhism as the religion of Tibet 1st Tibetan monks ordained by Śāntarakṣita Masters invited from abroad: Śāntigarbha Viśuddhasiṁha Jinamitra Dānaśīla Vimalamitra Translations underway on large scale 781 A.D. Monks invited from China 792–794 A.D. bSam-yas debate: Kamalaśīla, disciple of Śāntarakṣita, invited to Tibet	7th–8th c. Tantric themes in Khotanese and Kuchean art 8th c. Buddhist art of Karakhoto shows Tibetan influences 712 A.D. T'ang Emperor Hsüan-tsung favors Taoism 716 A.D. Pure Land teacher Ts'u-min returns from India 720 A.D. Vajrabodhi, Tantric master at Nālandā, travels to China 720–814 A.D. Po-chang Huai-hai, Ch'an master 736 A.D. Amoghavajra, disciple of Vajrabodhi, takes texts to China 763 A.D. Uighur rulers convert from Buddhism to Manicheanism 787 A.D. Tibetan Buddhist artists working at Tun-huang	736 A.D. Kegon teachings introduced into Japan by Indian monk Bodhisena 749 A.D. Bodhisena dedicates Buddha statue in Japan 752 A.D. Todai-ji temple of Kegon school built in Japan 778 A.D. Caṇḍi Kalasan temple built in Java c. 800 A.D. Caṇḍi Borobudur temple completed in Java 802 A.D. Founding of Hua-yen Haein-sa temple in Korea 802–854 A.D. Jayavarman II Cambodian king supports Mahāyana 805 A.D. Saichō returns to Japan from China with Shingon teachings 810 A.D. Kūkai brings Shingon teachings to Japan

CHAPTER EIGHTEEN

HEIGHT OF THE EMPIRE

During the reign of Khri-srong-lde-btsan, Tibet's political influence and military strength grew even greater. Old records show that Tibet expanded in all directions, subjugating Indian kings to the south, demanding taxes from T'ang China in the east, and again seizing the Silk Route in the north and west.

EXPEDITION TO INDIA

Sometime before bSam-yas was finished, a Tibetan army entered Magadha in northern India to obtain relics for the new temple. A marker was set up on the south bank of the Ganges, and a town erected for a military encampment. An unnamed Indian king is said to have surrendered and become a vassal of Tibet. No Indian records of this event seem to have survived, but the sBa-bzhed contains a full account.

The Indian ruler that Khri-srong-lde-btsan's army encountered must have been one of the Pāla kings, for northern India in the second half of the eighth century was ruled by the Pāla dynasty. According to Tāranātha, this dynasty began with King Gopāla, who ruled Bengal. He was followed by Devapāla,

Rasapāla, and Dharmapāla, whom Tāranātha calls a contemporary of Khri-srong-lde-btsan.

Modern histories, however, date Gopāla to about 750–770, Dharmapāla to about 770–810, and Devapāla to about 810–850. In any case, it seems accurate that Dharmapāla was contemporary with Khri-srong-lde-btsan and with Khri-srong-lde-btsan's son.

Since the Tibetan expedition took place before bSam-yas was finished, Tibetan troops must have entered India sometime between 762 and 775. The university of Odantapurī, which had been built by Gopāla, is mentioned in the sBa-bzhed's account. bSam-yas, already under construction, had been modeled upon Odantapurī, and so Gopāla had obviously already come to power.

The advance into India must have taken place during the reign of Gopāla's successor, Dharmapāla. Dharmapāla was a famous figure in Indian history, ruler not only of Bengal and Bihar, but also suzerain over the rulers of Kanauj, Avantī, Kuru, Gandhāra, and the Punjab. He founded the Buddhist university of Vikramaśīla and was patron and student of Haribhadra, who was a disciple of Śāntarakṣita. Since Śāntarakṣita, who helped found bSam-yas, had been abbot of Vikramaśīla before journeying to Tibet, it is clear that bSam-yas was erected during Dharmapāla's reign. Some Tibetan chronicles note that a king Dharmapāla was invited to Tibet by Padmasambhava and Khri-srong-lde-btsan, and arrived with an unusual statue of the Buddha carved in turquoise. Descendants of his were said to have settled in 'Phyongs-rgyas and included the family of the Fifth Dalai Lama.

Tibetan dominion in the south seems to have continued for several generations. An inscription at the tomb of Khri-srong-lde-btsan's son lDe-srong (Sad-na-legs) mentions an unnamed Indian king who surrendered to Tibet. The rGyal-po-bka'-thang puts the defeat of an Indian king named Dharmapāla in the reign of this same son (see chapter 19).

Though Tibet's power was strong in the south, she lost an ally in the north when the Western Turks were crushed by the Turkish Uighurs, a formidable power on the northern steppes after 745 A.D. The Uighurs often allied themselves with the T'ang Empire, but the Tibetans found new support among the Qarluq and Kirghiz Turks, who had moved into the old Western Turkish domains. This alliance, according to T'ang records, allowed trade caravans between the Arabs and Tibet to travel through the Qarluq territory in the far west.

CONQUESTS IN THE EAST

To the east, Tibetan forces successfully captured Tao-chou in the region of modern Co-ne in 755. The Tun-huang annals note that the minister Zhang mDo-bzher was then named general in charge of rMa-khrom, the military district that seems to have included the rMa-chu river region. It appears likely that this is the mDo-bzher-nag-po, the "ruler of eastern Tibet," with whom Vairotsana took refuge. In 754 the king of the tribes called Su-p'i by the T'ang historians was defeated by Tibet, and their crown prince Si-no-lo fled to T'ang territory. The Su-p'i lands, which extended to Mao-chou on the Min river and Ya-chou, were annexed by Tibet.

In 755 the T'ang Empire was badly shaken by the An Lu-shan military rebellion and other revolts that followed in its wake. Garrisons along the Silk Route were recalled to deal with the domestic troubles. This lack of military presence together with the general unrest greatly facilitated Tibet's expansion.

According to T'ang records, after 760, the Western Mountains of Chien-nan west of Ch'eng-tu, which had traditionally been the border of Tibetan and Ch'iang country, fell into Tibetan hands. By 763, Tibetan armies led by Zhang rGyal-zigs, Zhang sTong-rtsan, and Zhang sTag-sgra-klu-khong had marched five hundred miles southeast of Koko Nor. sTag-sgra-klu-khong was the loyal minister who had saved the life of Khri-srong-lde-btsan in 755, and he had become a trusted adviser and general. His forces now captured the T'ang capital at Ch'ang-an. They placed on the throne a brother of the Chinese princess Kim-sheng, the queen of Mes-ag-tshoms.

The invasion of the capital is recorded in the T'ang and Tun-huang annals, as well as on the inscribed pillar at Zhol in lHa-sa, where the reason for the invasion is explained: Upon the death of the emperor He'u 'Ki-Wang Te (Su-tsung r. 756–762), China had ceased to pay Tibet yearly tribute of 50,000 bolts of silk. The Tun-huang annals and the Deb-dmar also state that the tax silk had not been presented to the Tibetan court because of a change in China's government.

The Tibetan-supported emperor ruled only a few weeks, for the Tibetans were forced out of the capital. But it was several years before the T'ang forces could clear Tibetan troops from the area of the capital. This was finally accomplished in 765 with the help of 3000 Uighur cavalry. Tibetan troops continued an aggressive campaign. In 764 they took Liang-chou northwest of Lanchou. These two cities were on the route between Ch'ang-an and Tun-huang. Khri-srong-lde-btsan was obviously continuing the long-range strategy of his predecessors to seize the Silk Route.

In 774 an imperial T'ang decree ordered that their western borders be reinforced with over 230,000 troops. But Tibet struck to the north toward city-states at the east end of the Silk Route. These areas came under Tibetan control after the invasion of the T'ang capital. Kan-chou and Su-chou had fallen to Tibet by 766, Kua-chou was captured in 776, and Hami in 780. In 776 King Khri-srong-lde-btsan dispatched troops under Khri-sum-rje to take Tun-huang. After an eleven year siege, this crucial oasis state finally fell to Tibet in 787. Khri-srong-lde-btsan rewarded Khri-sum-rje by making him chief minister and commander of the army. After the capture of Tun-huang, the monk Mahāyāna (defender of the Chinese viewpoint in the bSam-yas debate), left for lHa-sa at the invitation of Khri-srong-lde-btsan.

RELATIONSHIP WITH T'ANG CHINA

In 781 at the beginning of the reign of Emperor Te-tsung (r. 780–805), who succeeded the emperors Su-tsung (r. 756–762) and Tai-tsung (r. 762–779), King Khri-srong-lde-btsan wrote to

Te-tsung, formally objecting to the mention of Tibet as "subject to China" in communications between the two countries. The Tibetan btsan-po's letter, as recorded in the T'ang annals, insisted that gifts sent to the T'ang court were not to be called tribute and that Tibet and China were equals allied by marriage. Te-tsung thereafter changed the language in his letters. The Tibetan king also insisted that the borders between China and Tibet be set in the Ho-lan-shan mountains (also known as the Ala-shan mountains), extending between Kan-chou and the present-day region of Ningxia. The emperor accepted this as well.

In 783 a treaty between T'ang China and Tibet was sworn at Ching-shui-hsien in the northeast, near the Lung-shan (Long-shan) mountains that separate the modern provinces of Gansu and Shaanxi. The treaty was inscribed on a pillar that was destroyed shortly thereafter, but its contents were also recorded in the T'ang annals. Tibetan territory was to extend up to Lan-chou, Wei-chou, Yuan-chou, and Hui-chou. All lands west of Lin-tao on the Klu-chu (Tao river), west of the Ta-tu river (rGyal-mo-dngul-chu), and west of the high mountains on the edge of the Ch'eng-tu plain belonged to Tibet. Land that was controlled by Tibet was to remain Tibet's, and what was controlled by China was to remain China's. The treaty was sworn in the traditional way with animal sacrifices, but at the suggestion of the Tibetan minister, Buddhist rites were also performed. The treaty was then also sworn again in the T'ang capital at Ch'ang-an.

During this period of formal peace, the T'ang annals note that Tibetan troops even helped the T'ang emperor in 784 against rebel forces near Ch'ang-an in return for a promise of additional territories, Ching-chou and Ling-chou. But when the lands were not given over, the fighting between the two armies began once again. Tibetan ministers planned revenge for the deceit. In 787 they ambushed a T'ang contingent at a peace conference at P'ing-liang; high-ranking T'ang officials were captured and sent to Khri-srong-lde-btsan. The T'ang annals note that Tibetan units were now encamped one after another near Feng-hsiang, only about 100 miles from Ch'ang-an. Tibetan soldiers disguised in T'ang armor raided a series of villages.

SETBACKS AND VICTORIES

Toward the end of the eighth century, Tibetan troops suffered several crushing defeats. The alliance with the Arabs was lost when a new ruler, Hārūn ar-Rashīd (r. 786–809), allied himself with China to halt Tibetan expansion. Hārūn ar-Rashīd is famous in the West as the Caliph of Baghdad in the *Tales of the Thousand and One Nights.*

The alliance with Nan-chao was also about to come to a treacherous end. According to the T'ang annals, 200,000 Nan-chao troops had joined 40,000 cavalry in 778 and attacked T'ang outposts along the Min river east of rGyal-mo-rong, but were badly defeated. I-meu-sin, son of Ko-lo-feng, who had been an ally of Mes-ag-tshoms, now joined his Nan-chao troops together with Tibetan forces to invade Ch'eng-tu in 781, according to Chinese sources on Yunnan history.

But in 793–94, I-meu-sin switched his allegiance to the T'ang Empire. Under the pretext of helping the Tibetan army regain the Silk Route, I-meu-sin had been amassing a large number of soldiers. These he sent to ambush the Tibetan troops in the Li-kiang region between Tibet and Nan-chao. As Tibetan forces crossed a suspension bridge over a river, the Nan-chao army cut the bridge and attacked, killing thousands of soldiers. Several southeastern frontier battles were now lost, and a Tibetan minister was captured. The Tibetan commander of troops in the southeast fled to China for fear of being demoted and punished. The T'ang records on Nan-chao mention that another battle was lost in the Nan-chao region in 801 when Arab and Sogdian troops commanded by a Tibetan general were forced to surrender.

Despite these setbacks in the southeast, Tibetan forces were very successful in the north. By 790 they had regained the Silk Route. An alliance of Tibetans, Qarluq Turks, Sha-t'o Turks, and White Turks defeated T'ang troops and their Uighur allies at Beshbaliq (Pei-ting), about seven hundred miles northwest of Koko Nor on the north branch of the Silk Route. The chief minister of the Uighurs, Il Ügäsi, attempted several times to regain the garrison with 50,000 troops, but without success. These battles in 790–791 are mentioned in Turkish inscriptions at Qara-Balghasun, the Uighur capital at Orkhon, as well as in the T'ang annals.

Tibet's victory secured its hold on the Tarim basin, and Tibetan military supervision was established over local kings, who were permitted to retain their thrones. Tibet's control of Central Asia lasted into the ninth and tenth centuries. Several centuries of later Turkish control were ended when the region became part of the Mongol Empire in the thirteenth century.

CENTRAL ASIAN EMPIRE

The expansion of the Tibetan Empire into Central Asia is foretold in prophecies in the canonical texts on Khotan such as the Li'i-yul-lung-bstan-pa, the Dri-ma-med-pa'i-'od-kyi-zhus-pa, and the Ri-glang-ru-lung-bstan, which refer to the days when the Tibetans will seize Li-yul.

Evidence of the huge extent of the empire can be found in an inscription from the pillar at the bridge in 'Phyongs-rgyas, carved at the time of Khri-srong-lde-btsan. Calling him the Divine King beyond comparison with other kings in the four directions, the inscription notes that his dominion extended from the borders of Ta-zig (Arab and Persian lands) in the west to the passes at Long-shan in the east. The Tun-huang annals also note that Khri-srong-lde-btsan's empire extended to Long-shan.

Evidence of Tibetan occupation of Central Asia also includes Tibetan documents from Tun-huang, as well as from sites between Khotan and Tun-huang along the trade routes. In 1906 a Tibetan fort was excavated near Lop Nor at Miran. In addition to Tibetan records, lacquered leather armor, arrows, and Tibetan seals made from animal horn were found. A huge collection of documents was discovered at another fort at Mazar Tagh just north of Khotan, including military reports and requisitions. This fort lay along the Khotan river and guarded the routes connecting the southern branch of the Silk Route that ran through Khotan with the northern branch that ran through Kucha.

Sometime after the Silk Route was secured, the Tibetan army attacked the Uighurs known as the Bhaṭa Hor, whom R. A. Stein locates in the Kan-chou region. Their goal was to bring the guardian deity Pe-har to bSam-yas. Some sources give the credit for the victory to the general Klu-dpal together with Khri-srong-lde-btsan's middle son Mu-tig.

CULTURAL CONTACTS WITH OTHER LANDS

The Central Asian lands, a meeting ground of eastern and western civilization for many centuries, must have offered Tibet new inspiration of many kinds. In the period since the beginning of the Christian era, the oasis states of Central Asia had become high civilizations with cosmopolitan artistic and religious traditions reflecting influences from Greece, Persia, and India as well as from China and the Turks. But Tibet's cultural contacts in the far north and west are not as well documented as the military conquests.

We do know that Khotanese artists and craftsmen had been working in Tibet, and that Tibetan teachers such as Vairotsana had also made connections with the older, more established Buddhist communities in Khotan. Buddhist texts were being translated at this time from a number of different languages, some of which were Central Asian. The full extent of the relations with Khotan remain to be investigated.

What cultural influence came from the Turks is also not fully known, but the Tibetan-Turkish alliance included several marriages, and some cultural exchange must have taken place.

Trade with the Arabs is noted in the *Ḥudūd al-'Ālam,* which says that "Tubbat" imported many items from India and then exported them to Muslim lands, together with their own products such as animal furs, gold, and musk. The *Tārīkh al-Ya'qūbī* mentions that on one occasion Tibet gave the Arab Caliph, al-Ma'mūn, a golden statue, which was then housed in Mecca.

Other Western influences reached Tibet in King Khri-srong-lde-btsan's time. By the eighth century Nestorian Christian and Manichean teachings had both become popular in Sogdia, and from there had spread across Central Asia. The king himself mentions Mani in one of his expositions on religion. Nestorian texts in Chinese have been found at Tun-huang, and Christian material in Sogdian and Turkish translations has been discovered at various Central Asian sites.

To determine how widely known Christian teachings might have been in Tibet will require more research. But that there was some contact is quite clear. A Tibetan book of divination found at Tun-huang mentions "Jesus, the Messiah." Crosses carved in rock, together with nearby inscriptions in Sogdian, have been noted in Ladakh. An interesting letter written in the eighth century by the leader of the Nestorian church, the Patriarch Timothy I (r. 780–823), describes his intention to send a Nestorian primate into Tibet.

There was cultural contact also with Persia. The design of eighth century Tibetan armor seems to have been borrowed from Persia. Tibetan armor was remarked upon by T'ang historians, who greatly admired it. Finely made of small, overlapping plates of metal, with openings only for the eyes, this armor was flexible, and yet so strong no arrow could pierce it. Just such armor was used by the Persians, according to scholars who compared these descriptions with sculptures of Persian kings and descriptions of Persian armor given by Roman historians.

A Persian physician, Halaśanti, arrived in Tibet at the invitation of Khri-srong-lde-btsan. According to the Shes-bya-kun-khyab and the mKhas-pa'i-dga'-ston, he translated western medical texts and possibly Turkish and Sogdian ones as well. The king also invited Indian and Chinese doctors, and the three foreign physicians worked together to compile a medical sourcebook. The renowned Tibetan physician, gYu-thog-mgon-po (708–833), traveled to India many times to study medicine and collect texts. The Indian traditions of medicine, literature, and grammar were also entering Tibet together with Buddhist teachings.

Perhaps because of rapid changes taking place in Tibetan society, Khri-srong-lde-btsan saw the need to set forth new social laws for the people, enlarging on the code devised by Srong-btsan-sgam-po. The Tun-huang documents mention a great code that prescribed proper punishments and compensations, and was applied with justice and diligence. Other histories state that regulations were established for payments of debts, loans, and wages, and specific punishments were set for crimes. Guidelines were laid down for treatment of the sick and the poor. People were encouraged to learn to read and to study writing and arithmetic, while teachers were urged to educate all the children in reading, writing, and spelling. This code was put into writing, and Khri-srong-lde-btsan specified that it be read to every single Tibetan at least once.

KHRI-SRONG-LDE-BTSAN'S HEIRS AND HIS RETIREMENT

To further unite the Tibetan Empire from within, Khri-srong-lde-btsan married five wives from among the powerful families of Tibet: lHa-mo-btsan from the mChims clan; Byang-chub-sgron from the 'Bro clan; rGyal-mo-btsun from the Pho-yong family; rMa-rgyal mTsho-skar-ma from the Tshe-spong clan (known in some sources as dMar-rgyan); and Ye-shes-mtsho-rgyal from the mKhar-chen family. mTsho-rgyal became one of the leading disciples of the Tantric master Padmasambhava, while rMa-rgyal became the mother of Khri-srong-lde-btsan's son Mu-ne and his youngest son, lDe-srong. lDe-srong is also known as Sad-na-legs and is called Mu-tig by some sources. A third son, known sometimes as Mu-rug or Mu-tig, is not mentioned in the Tun-huang documents, but is known to later histories and is mentioned as "Mu-rug, the elder brother of lDe-srong" in the edict on the pillar at Zhwa'i-lha-khang.

According to the sBa-bzhed, sometime after Vimalamitra arrived in Tibet, Khri-srong-lde-btsan retired, and his son Mu-ne took the throne. But disaster befell the middle son, who was expelled for killing a minister. Having been dismissed from a conference of the great ministers, Prince Mu-tig in his anger had cut down one of the king's advisers with his sword. As punishment, he was exiled to the Mon country in the south at lHo-brag near Bhutan, though some sources say he was sent to the northern frontiers.

Tibetan historians have preserved several different accounts of the date of Khri-srong-lde-btsan's death, the length of his life, and the reign of Mu-ne. Perhaps this is a reflection of the uncertainty and power struggles that arose after Khri-srong-lde-btsan's reign. According to the sBa-bzhed, Khri-srong-lde-btsan did not die until sometime after his son Mu-ne took the throne. The Deb-dmar notes that Khri-srong-lde-btsan was still alive even after Mu-ne died. This source puts Mu-ne's reign of a year and nine months in 786–787 and the king's death shortly thereafter.

The Deb-dmar also records a different chronology based on T'ang sources that states Khri-srong-lde-btsan died in the iron-monkey year of 780, and Mu-ne ruled for seventeen years until the fire-ox year of 797. Another king Ju-tse ruled til 804 when lDe-srong became king. The Blue Annals quotes this chronology from the Deb-dmar, but we have not been able to find Ju-tse listed in any other source. Western scholars note a confusion in the names and dates of Tibetan rulers given in the T'ang annals for the last decades of the eighth century. This may have misled the Tibetan historians who referred to Chinese sources.

We can be confident that Khri-srong-lde-btsan was ruling as late as 794 because he presided over the bSam-yas debate, which lasted from 792 until 794. Thus if a king died in 797, it would have been Khri-srong-lde-btsan. This would make Mu-ne's rule much shorter, which fits with a number of accounts stating that he ruled only about a year and a half. If this is the case, then Khri-srong-lde-btsan's reign lasted forty-two years, and he was fifty-six when he passed away in 797. The tomb of King Khri-srong-lde-btsan was erected in 'Phyongs-rgyas near that of his father, Mes-ag-tshoms.

THE ERA OF KHRI-SRONG-LDE-BTSAN

During Khri-srong-lde-btsan's reign, Tibetan literature, art, and architecture blossomed, inspired by the Buddhist tradition and cultural contacts with other lands. The Tibetan Empire was at its height, controlling Khotan and the coveted Silk Route in the north, reaching the Turkish and Arab lands in the west, and stretching to Lan-chou and the rGyal-mo-rong river in the east, even threatening the T'ang capital at Ch'ang-an. Tibetan histories sometimes claim that Khri-srong-lde-btsan controlled one-third of Asia. During the last few decades of the eighth century the Tibetan Empire appears to have been one of the three greatest powers in Asia.

BORDERLANDS WITH T'ANG CHINA

During the reigns of Khri-srong-lde-btsan and T'ang Emperor Te-tsung, a new peace treaty was agreed upon. This treaty was sworn in 783 at Ching-shui-hsien near the Lung-shan mountains. The contents were recorded in the T'ang annals, which note that Tibetan territory was to extend to Lan-chou, Wei-chou, Yuan-chou, and Hui-chou; all lands west of Lin-tao on the Tao river (Klu-chu), west of the Ta-tu river, and west of the mountains of Chien-nan belonged to Tibet. Armies were to cease hostilities and no further incursions were to take place.

CONTROL OF CENTRAL ASIA

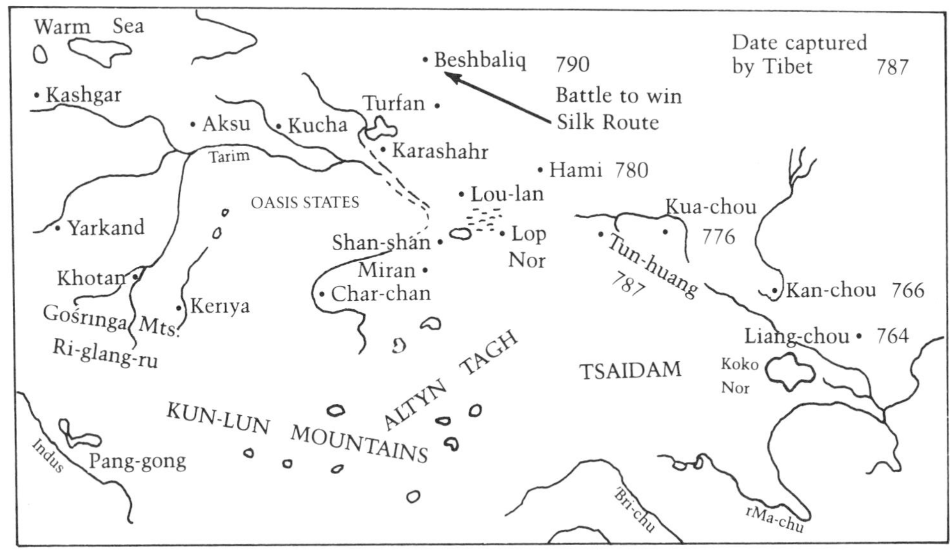

CENTRAL ASIAN PLACE-NAMES

CENTRAL ASIAN	OLD CHINESE	PINYIN
Kashgar	Su-lo	Sule
Yarkand	So-chu	Shache
Kucha	An-hsi	Kuqa
Aksu	Ku-mo	—
Karashahr	Agni/Yen-ch'i	Yanqi
Turfan	Kao-ch'ang	Turpan
Khotan	Yu-tien	Hotan
Beshbaliq	Pei-t'ing	Beiting
Charchan	Chu-mo	Qiema
Keriya	Han-mo	Yutian
—	Tun-huang	Dunhuang
—	Kan-chou/Chang-yeh	Zhangye
—	Su-chou	Jiuquan
—	Liang-chou	Wuwei

EIGHTH CENTURY EMPIRE

In comparison with other kingdoms and dynasties at the end of the eighth century, Tibet was one of the most powerful in Asia during the reign of Khri-srong-lde-btsan. T'ang power had begun to decline since the 750s, while the Arab armies could expand no farther eastward. The Turks were fragmented; Byzantium was a small shadow of the former Roman Empire; and the Pāla kings and the Kārkoṭa dynasty controlled only the northern parts of India.

SOUTHEASTERN TRADE ROUTES

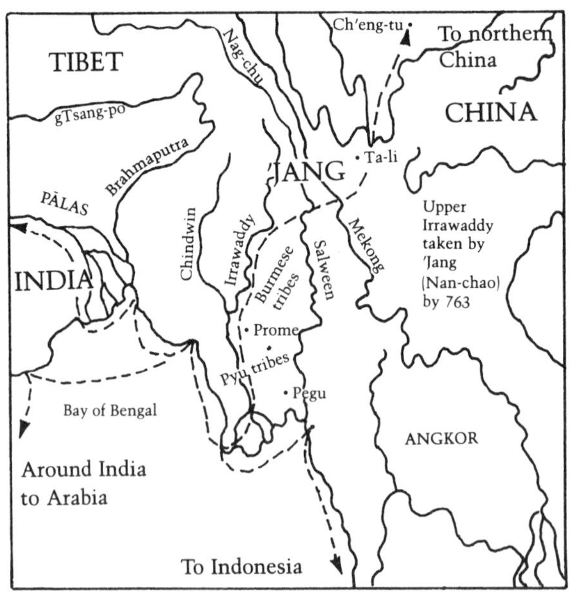

Trade routes connecting India, Burma, and China ran through 'Jang (Nan-chao), making it territory that was contested between the Tibetan army and the T'ang army. In 678 tribes around Lake Tali in Nan-chao submitted to Tibet. By 703 during the reign of 'Dus-srong, Nan-chao peoples known as the Black and White Mya-ba became Tibetan vassals. In 751 Ko-lo-feng and Mes-ag-tshoms became allies. This relationship lasted until the end of the eighth century.

THE ERA OF KHRI-SRONG-LDE-BTSAN

TIBET	MIDDLE EAST	FAR EAST	EUROPE
745 A.D. Tibetan allies, W. Turks, crushed by Uighurs	739 A.D. Arabs advance into Central Asia	c. 740–760 A.D. Nāgabhata I rules N. India, defeats Arabs	c. 750 A.D. Byzantine Empire controls southeast Europe and Turkey
756–797 A.D. Reign of Khri-srong-lde-btsan	c. 750 A.D. Turkish Khazars living north of Caspian Sea	755 A.D. An Lu-shan rebellion in China: imperial authority declines	755 A.D. Successive wars between Bulgars and Byzantium begin
763 A.D. Tibetan troops invade Chinese capital at Ch'ang-an	c. 750 A.D. Under Abbasid Caliphs, power in Islamic Empire shifts to non-Arab majority; Persian influence increases	745–840 A.D. Uighur Turks establish Orkhon Empire in Mongolia	756 A.D. Muslim dynasty established in Cordova, Spain; refuses to recognize Abbasid rule
763 A.D. After foundation of bSam-yas, Tibetan troops invade N. India	c. 760 A.D. Arabs adopt Indian numerals and develop algebra	752 A.D. Todai-ji temple built in Japan	787 A.D. Christian council at Nicaea allows worship of religious images; reversed in 815 at council of St. Sophia
766 A.D. Tibet takes Kan-chou	763 A.D. Abbasid capital moved from Damascus to Baghdad	763 A.D. Uighurs convert to Manicheanism	793 A.D. Vikings raid N.W. Europe
776 A.D. Kua-chou captured by Tibet	786–809 A.D. Caliph Hārūn ar-Rashīd reigns; Golden Age of Islamic learning begins	766–801 A.D. First Chinese historical encyclopedia compiled	800 A.D. Charlemagne, Frankish King, crowned Emperor of Rome; controls most of W. Europe
780 A.D. Hami taken by Tibet	788 A.D. Abbasids begin to lose control of distant parts of Empire: Tunisia, Morocco, east Persia	780 A.D. Korean kingdom of Silla declines	804–806 A.D. Arab raids on Byzantium
783 A.D. Treaty with China sworn at Ching-shui	813–833 A.D. Caliph al-Ma'mūn supports arts and sciences, translations from Greek, Sanskrit, and Persian	794 A.D. Japanese capital changed from Nara to Kyoto	807 A.D. War between Franks and Byzantium
787 A.D. Tibet takes Tun-huang		c. 800 A.D. Completion of Borobudur temple in Java	814 A.D. Disintegration of Charlemagne's Empire
789 A.D. Tibetan-Arab alliance ends		802 A.D. Jayavarman II founds Angkor Kingdom in Cambodia	
790 – c. 850 A.D. Tibet controls Khotan, Silk Route			

CHAPTER NINETEEN

SAD-NA-LEGS AND RAL-PA-CAN

Upon the retirement of Khri-srong-lde-btsan, a struggle seems to have broken out among the sons and wives of the great king. After royal command passed out of the hands of one of the most influential rulers in Asia, the Tibetan throne was occupied by two successive heirs in just a few years.

Khri-srong-lde-btsan's son Mu-ne was first enthroned. In a brief reign that probably lasted only about a year and a half, he attempted on three occasions to equalize the wealth of the land, distributing riches to the poor. Even though he did not succeed in bringing about any permanent change, these measures may have contributed to his downfall by antagonizing wealthy nobles.

Khri-srong-lde-btsan had insisted Mu-ne marry one of his young queens, mDo-rgyal of the Pho-yong family, apparently to protect her from the jealousy of his other wives. But soon both Mu-ne and his wife were murdered by Mu-ne's mother, the Tshe-spong queen of Khri-srong-lde-btsan. After Mu-ne's death, the third son lDe-srong was made king. The middle son, who had been exiled, was killed by the sNa-nam clan, but apparently not until after lDe-srong became king. In an inscription at Zhwa'i-lha-khang, lDe-srong notes that he took power after his father and his brother had died, and that he bound his older brother with an oath.

Clarifying the events surrounding Mu-ne-btsan-po's reign will require further research, but it seems likely that Mu-ne ruled from 797 to 799 and that lDe-srong's rule began sometime in 799 or 800. Despite the confusion surrounding the final years of the ninth century in Tibet, it is clear from T'ang and Tibetan sources that by 804 lDe-srong was on the throne. Though his ruling title was Khri-lde-srong-btsan, he is more commonly known as Sad-na-legs.

SAD-NA-LEGS SUPPORTS BUDDHISM

The young king was a supporter of the Dharma like his father before him and had been a disciple of Padmasambhava since childhood. Several sources state that the Great Guru left Tibet in the reign of Sad-na-legs. Though there are several accounts about how long Padmasambhava worked in Tibet, a number of them note that he left in a monkey year. This might well be 804, thus making his stay forty-two years long. If he stayed until 816, the next monkey year, the reign of King Ral-pa-can would have already begun.

During the reign of Sad-na-legs, a number of Dharma masters who had arrived in the time of his father continued to be active, and the king supported translations and the building of temples, such as sKar-chung-rgya-sde. In an edict carved in stone and set up outside this temple, the king again proclaimed Buddhism as the religion of Tibet, just as his father had done, and all the queens and ministers swore an oath of loyalty to the Dharma.

Sad-na-legs relied greatly upon the advice of his Buddhist teachers and even gave them political appointments. Myang Ting-nge-'dzin, a disciple of Vimalamitra, had been the guardian and teacher of the young Sad-na-legs. Deeply devoted to the king, Myang offered him advice and support throughout his reign, becoming an influential minister in the government. An inscription at Zhwa'i-lha-khang, a temple built by Myang, recounts the gifts of land and privileges given to this minister and thanks him for the great benefit he brought the Tibetan people.

Minister dPal-chen-po, a Buddhist monk known as Bran-ka dPal-gyi-yon-tan, was also active in foreign affairs during the reigns of Sad-na-legs and his son. Letters to dPal-chen-po from the emperor of China, Hsien-tsung (r. 805–820), written about 810, asked the monk to work for peace between their countries. The emperor urged dPal-chen-po as chief minister to make a request to the Tibetan king for the return of three districts taken by the Tibetan army, Ch'in, Yuan, and An-lo southeast of modern Gansu, and for the release of prisoners of war. No land was returned, and a peace treaty was not signed for another decade. A similar letter was written to the Tibetan minister Khri-sum-rje. The letters from the emperor are preserved in the collected works of a ninth century T'ang author.

THE NINTH CENTURY TIBETAN EMPIRE

Sad-na-legs controlled a vast territory that included all the conquests of the earlier kings: the whole of the Tibetan plateau to the Lung-shan mountains and the rGyal-mo-rong river in the east, as well as the Central Asian oasis states. Eighth and ninth century Tibetan documents from Central Asia reveal the names of the military divisions of the frontier lands of the empire. Each frontier zone, known as a khrom, was headed by a dmag-dpon or military commander. According to Uray's research, rMa-khrom included the rMa-chu area south of Koko Nor; dByar-mo-thang-khrom was to the northeast; mKhar-tsan-khrom was probably the Liang-chou region; Kwa-chu-khrom included the Tun-huang area; Tshal-byi was probably the Lop Nor region along the south side of the Silk Route; Bru-sha-yul-gyi-khrom was the Gilgit area; Khotan was generally known as Li-yul.

These regions were much larger than T'ang administrative districts, which were called chou. In regions where chou had been established, such as Tun-huang and Kua-chou, several old chou were included within one khrom. Documents from Tun-huang show that the organization and administration of a khrom involved at least forty ranks of officials including a war cabinet and ministers in charge of agriculture, fortifications, inspections, and taxes. To run these huge regions smoothly, judges, census takers, town prefects, and numerous scribes were also required.

The internal administration of central Tibet was separate from the military governments of the frontiers. dBus and gTsang were divided into five large "wings" with one in the center, one to the right and one to the left, to which two additional wings were added. dBu-ru, the central wing, was around the lHa-sa region. The wings faced south so that the left wing called gYo-ru was in eastern dBus, and the right wing of gYas-ru was to the west in gTsang. The wing of Ru-lag was also in gTsang, east of gYas-ru. The wing called Sum-pa'i-ru seems to have included the Sum-pa tribes around Nag-chu-kha. Powerful clans were in charge of regions within the wings. Each region of Tibet provided soldiers for the army, which, according to the figures in the Blon-po-bka'-thang, was nearly three million men strong. It is clear from records such as these that the population of Tibet in the eighth and ninth centuries was greater than in recent times.

FOREIGN RELATIONS IN THE REIGN OF SAD-NA-LEGS

Sad-na-legs' reign is known in less detail than those of his predecessors, for the year-by-year court annals from Tun-huang extend only up to 763 A.D., and the Tun-huang chronicles end with Khri-srong-lde-btsan's reign. A number of other early sources, however, agree that Sad-na-legs made treaties or agreements with rulers in the four directions.

There was no formal treaty with T'ang China during his reign. T'ang records show battles at Yen-chou, Lin-chou, Wei-chou, Ya-chou, and Sui-chou between 800 and 803. Thus, fighting continued in the borderlands east of Koko Nor and along the Ta-tu river (rGyal-mo-dngul-chu). But from 804 onward, envoys traveled regularly between lHa-sa and Ch'ang-an, according to T'ang reports. When the Chinese Emperor Te-tsung died in 805, Sad-na-legs sent gifts of gold, silver, cloth, and domestic animals to the funeral.

A description of this king's diplomacy can be found in the short historical summary at the opening of the treaty between T'ang China and Tibet made by the next king and inscribed on a stone pillar at lHa-sa. There it is stated that both Sad-na-legs and the T'ang emperor desired peace and conferred about a treaty, but no agreement was made in his reign. The inscription also notes that Sad-na-legs made agreements with the other kings in the four directions, but no names of countries or kings are given.

More information can be found in the inscription erected at the tomb of Sad-na-legs. It begins by comparing him, like his father before him, to the first king of Tibet, a god from heaven come to rule men. Sad-na-legs is described as a man with a firm command of the government. His spirit was profound, his heart was magnanimous, and his mind keen and comprehending.

The inscription then praises his political dealings with the rulers in the four directions. Although the stone is badly damaged and the writing difficult to decipher, a copy made by Kaḥ-thog master Tshe-dbang-nor-bu in the eighteenth century reveals some of the details. Campaigns were undertaken against "Upper China" as soon as Sad-na-legs took power, for disagreements had arisen between the two countries. But eventually peace was made. In India there were dealings with a great king, but no name is detectable on the inscription. The Hor (Uighurs) and the Dru-gu (Turks) are mentioned, but the lines are incomplete.

The rGyal-po-bka'-thang has a very similar passage describing how Sad-na-legs extended his power in four directions, defeating China to the east, King Dharmapāla and King Dra'u-dpun in the south, Ta-zig kings La-mer-mu and Hab-gdal in the west, and Dru-gu (Turk or Uighur) and Li-yul (Khotan) kings in the north.

THE IDENTITY OF THE FOREIGN KINGS

Though these sources all agree that treaties were made with kings in the four directions, it is difficult to determine the identities of the kings mentioned in the rGyal-po-bka'-thang. The Uighurs were enemies of Tibet in the 8th and 9th centuries, helping drive Tibetan troops from the region of the T'ang capital in 763–765. Later they were defeated by Tibet in 790 when the Silk Route was regained. T'ang annals mention 50,000 Tibetan soldiers attacking the Hui-ho (Uighurs) in 806, but we have not been able to find reference to any treaty with the Uighurs during the reign of Sad-na-legs (but see chapter 20 for a later treaty with the Uighurs).

Khotan and the Silk Route remained under Tibetan control in the reign of Sad-na-legs, but no specifics on early ninth century battles or treaties with Khotan could be found in other contemporary sources such as the T'ang annals. This is not surprising, for China had lost all contact with Khotan and the other oasis states, and the T'ang annals have nothing further to say about Central Asia after the Tibetan army took control of the Silk Route in 790.

The identity of King Dra'u-dpun is obscure, but King Dharmapāla in the south certainly seems to be the Pāla ruler of northern India, who reigned perhaps between 770 and 810, and thus might be contemporary with three different Tibetan rulers: Khri-srong-lde-btsan, Mu-ne-btsan-po, and Sad-na-legs.

Indian records do not mention any struggles with Tibet, but the Arab records of this era are more helpful. Tibetan troops joined the Qarluq Turks in 809 to assist rebels in the Sogdian capital of Samarkand against the Arab Caliphate, according to the *Tārīkh al-Ya'qūbī.* al-Azraqī reports in a ninth century work that the Shah of Kābul, whose lands were to the west of Oḍḍiyāna, had become an ally of Tibet at an uncertain date, and was known to the Arabs as a "king among the kings of Tibet." But in 812–13 he submitted to the Arabs, and the Tibetan armies were defeated in several struggles with the Arabs in the west.

Within the next few years, however, there seems to have been peace between Tibet and the Arab Caliphate, for the *Tārīkh* mentions that a Tibetan official made a visit to the Arab Caliph al-Ma'mūn (r. 813–833), Hārūn ar-Rashīd's son and successor in Khurasan (the Arab territory in western Central Asia and Persia). It seems possible that the La-mer-mu of the rGyal-po-bka'-thang could be al-Ma'mūn. But the identity of the other Ta-zig king Hab-gdal is still a mystery.

RAL-PA-CAN'S REIGN BEGINS

Khri-lde-srong-btsan Sad-na-legs died sometime about 815, after a reign of about fifteen years. The Blue Annals gives a wood-horse year of 814 for his death, while the T'ang annals record a fire-bird year of 817, which may have been the year the T'ang court was informed. But early Sa-skya sources also say he died in a fire-bird year.

Sad-na-legs had married four wives: lHa-rtse, Legs-mo-brtsan from the mChims clan, brTan-rgyal of the Cog-ro clan, and lHa-rgyal of the 'Bro clan. He had five sons, lHa-rje, lHun-grub, gTsang-ma, Ral-pa-can, and 'U-'i-dum-brtan. Ral-pa-can, who had been born in 806, became the next king at age eleven or twelve.

Raised by a father who was a devoted supporter of Buddhism and surrounded by Buddhist teachers in the court, Ral-pa-can, whose reigning title was Khri-gtsug-lde-btsan, became a renowned Dharma King. Though earlier kings such as Srong-btsan-sgam-po had laid the foundation for Buddhism very quietly, Ral-pa-can proclaimed his faith in the Buddhist teachings every day in open court. He applied the principles of the Dharma to the running of his kingdom, and the Buddhist tradition considers him an incarnation of Vajrapāṇi.

The king remodeled older temples and near the confluence of the gTsang-po and sKyid-chu rivers built the nine-storied temple of 'U-shang-rdo, famous for its remarkable golden roof. Some sources report that up through the reign of this king, 1008 temples had been erected by Tibetan kings in China and Tibet. Artisans worked extensively on Buddhist temples, but the royal summer camp near lHa-sa, a palatial military pavilion, was also wonderfully decorated with golden figures of tigers, panthers, and dragons. Beautiful statues in gold and silver were also sent as gifts to foreign lands.

Ral-pa-can invited craftsmen from China, Nepal, and Kashmir, but he especially wanted the artists from Khotan. The king even threatened Khotan with a Tibetan army if it hesitated to dispatch the desired "King of Artisans" with his three sons to work in Tibet.

RAL-PA-CAN WORKS FOR THE DHARMA

With great foresight, Ral-pa-can realized that the successful transmission of the written Dharma to later generations would require developing commonly accepted terms with completely clear and regular meanings. Since translations in earlier times had been made from texts in various languages and had not used a standardized terminology, the king established a commission to revise these translations. Buddhist masters, including Jinamitra, Dānaśīla, Surendrabodhi, Śīlendrabodhi, and Bodhimitra worked together with Tibetan teachers and skilled translators, including Ye-shes-sde and sKa-ba dPal-brtsegs. The staff recast translations, regularizing the vocabulary, and compiled a large lexicon, the Mahāvyutpatti, which listed the accepted Tibetan equivalents for thousands of Sanskrit terms.

Ral-pa-can gave extensive privileges to the Buddhist monks, and honored teachers and monks in all situations. Laws were established assigning the responsibility for the upkeep of each Buddhist monk to seven families, and strict punishments were meted out to anyone disrespectful toward monks. Such displays of appreciation for the Dharma, however, stirred up increasing jealousy and resentment among some of the old noble families and Bon supporters.

Stories are told of how this king, ruler of one of the most extensive empires in Asia, would seat the monastic Sangha upon silken scarves attached to his own hair as an example of the proper respect owed the Dharma. Naturally, the Buddhist ministers of state grew even more powerful during his reign.

RAL-PA-CAN'S EMPIRE

During his reign of some twenty years, Ral-pa-can continued to preside over the territories in Central Asia, and is reported by some later sources to have made conquests to the south that included Mon, Blo-bo, Za-hor, and some part of India, reaching even to the mouth of the Ganges in Bengal. Tenth century Arab authors such as ibn-Haukal were still referring to the Bay of Bengal as the Tibetan Sea, indicating a strong and persistent Tibetan influence far to the south.

Though modern Indian historians mention the Tibetan records of conquests in India from the days of the Dharma Kings, they have not yet found any Indian records of relations with Tibet. There is no mention of Srong-btsan-sgam-po's invasion, though it is documented in the T'ang annals; nor of Khri-srong-lde-btsan's invasion reported by the sBa-bzhed; nor of dealings between Sad-na-legs and an Indian king, inscribed on a pillar and recorded in the bKa'-thang. Recalling the strong religious and cultural ties forged between Tibet and India in the eighth and ninth centuries might inspire researchers to continue the search for fuller documentation of military and political events in Indian historical sources.

THE 822 TREATY WITH T'ANG CHINA

The greatest political accomplishment of Ral-pa-can's reign was the famous 822 treaty with T'ang China. In 821 Ral-pa-can sent ambassadors to Ch'ang-an with the text of a treaty that he was prepared to agree to. The Emperor Mu-tsung concurred, and the agreement was signed and witnessed first in Ch'ang-an. In 822 T'ang officials were dispatched to Tibet for a second signing ceremony at the Tibetan king's summer residence south of lHa-sa. The T'ang annals contain a description of the ceremony which was attended by over one hundred officials and was sworn with both Bon and Buddhist rites at an altar near the king's royal tent.

The full text of the treaty can be seen on an inscribed pillar set up at the Jo-khang in lHa-sa in 822. On the east face is an historical account of the alliances between China and Tibet. The west face contains the treaty itself in both Chinese and Tibetan. The south face lists T'ang ministers who signed, and the north face gives the names of the Tibetan officials who signed; both these faces are bilingual as well.

The historical account names the first Tibetan king, a god who descended to be king of men, a marvelous hero and great military leader who established the Tibetan laws. In the time following his reign, Tibet's power grew until all its neighbors revered Tibet — Persia, India, the Turks. Compared to its relations with other countries, such as Nepal, Tibet had a special relationship with China, a land whose "wisdom and customs were a match even for Tibet."

The rest of the east face recalls alliances with China: the marriages of two princesses to Tibetan kings Srong-btsan-sgam-po and Khri-lde-gtsug-btsan. Interestingly, Khri-srong-lde-btsan, who badly defeated the T'ang armies many times, is not named, even though a peace treaty was concluded during his reign in 783. The west face with the actual treaty explains how the great king of Tibet, the divine manifestation, the Tibetan btsan-po, and the great king of China, the Chinese ruler Huang Te (Mu-tsung r. 820–824), as son-in-law and father-in-law, made an alliance. With "great profundity of mind" and concern for the future as well as for the present, the two rulers renewed their friendship and ceased hostilities. They jointly declared that both Tibet and China would keep the territories and frontiers they possessed at that time, with the whole region to the east of the frontier being the country of Great China and the whole region to the west being the country of Great Tibet.

This treaty did not describe boundaries between Tibet and T'ang China, which had already been laid out in the 783 treaty in detail. The short summary in the T'ang annals simply repeats that the T'ang were to be sovereigns over the lands under their control at the time of the treaty, while in the west Great Tibet was to be the master. By the time of the 822 treaty, the Tibetan

Empire was even greater than in 783, including the lands of Khotan and the rest of the Tarim basin, as well as Tun-huang and regions in the northeast.

For the next fifteen years, ambassadors traveled regularly between lHa-sa and Ch'ang-an, and relations appear to have been peaceful. Tibet requested a map of the mountain Wu-t'ai-shan and sent statues of animals cast in silver as gifts to the T'ang court.

Shortly after the peace of 822 a monastery named dGe-ba-g·yu-tshal was founded in the northeast in the frontier zone of dByar-mo-thang. The ministers of various regions offered donations together with five sets of beautiful, poetic prayers, which were found among the records at Tun-huang. These prayers praised the Tibetan kings and particularly honored Khri-gtsug-lde-btsan (Ral-pa-can), signer of the peace treaty. This great king and his ministers were held in awe by the frontier cities because they "protected the people, tamed the borderlands, and humbled all Tibet's enemies."

Texts of these prayers also mention that treaties were made at this time not only with China, but also with Nan-chao ('Jang) and the Uighurs (Drug), both sometime allies of the T'ang. Sa-skya historians note that a treaty with the Uighurs was made in 822; T'ang annals likewise mention this treaty and add that a marriage alliance was arranged as well.

In the days of the last two Dharma Kings, all areas of cultural life were stimulated—art, architecture, literature, religion, technology. Thus as the political power of the empire grew, so too did its cultural achievements. Though the empire was to decline, these cultural advances were incorporated into Tibetan civilization, to be preserved for over a thousand years.

Chos-rgyal Ral-pa-can

TERRITORIAL DIVISIONS OF THE EMPIRE

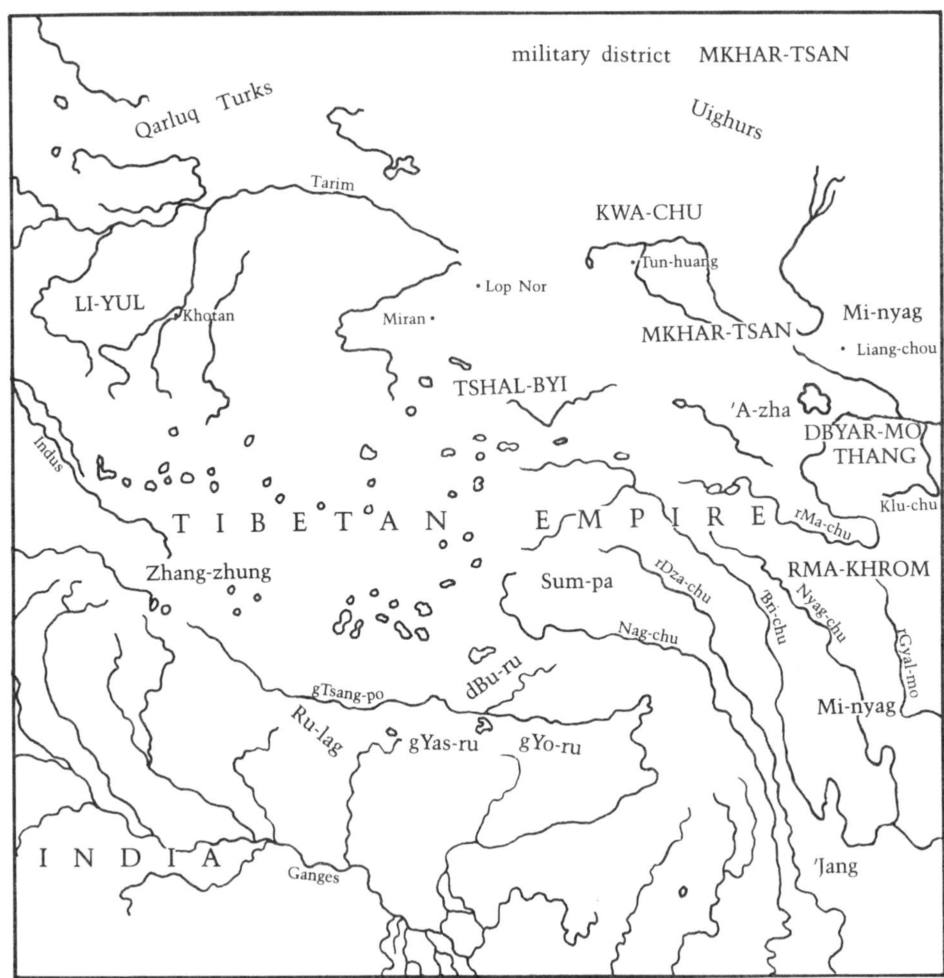

By the ninth century the Tibetan Empire was divided into khrom, territories that comprised conquered regions in Central Asia and the northeast. The khrom were administered separately from the internal divisions of Tibet, known as ru: dBu-ru, gYas-ru, gYo-ru, Sum-pa'i-ru, and Ru-lag. The khrom had their own administration that required over forty ranks of officials. Tibetan forts in Central Asia, such as those discovered at Miran and Mazar-tagh, defended these regions and served as outposts for military expeditions.

THE ERA OF SAD-NA-LAGS AND RAL-PA-CAN

TIBET	INDIA, KASHMIR	FAR EAST	EUROPE
797–800 ? A.D. Mu-ne-btsan-po rules, attempts to equalize wealth	750–770 A.D. Gopāla rules Bengal, elected as king by popular vote; builds university of Odantapurī	794 A.D. Heian Period begins in Japan	800–814 A.D. Frankish King Charlemagne rules W. Europe (W. Roman Empire)
800–814 A.D. Sad-na-legs rules, supports Dharma	757–800 A.D. Kṛṣṇa, greatest king of Rāṣṭrakūṭa dynasty in S. India	802 A.D. Jayavarman II establishes Angkor Cambodian kingdom	c. 800 A.D. Western Moslems terrorize Mediterranean
Treaties or agreements with China, Turks, India, Persia	770–810 A.D. Dharmapāla rules Bengal and Bihar; suzerain over Kanauj, Avantī, Kuru, Punjab, and Gandhāra	805–808 A.D. Tendai and Shingon Buddhist sects founded in Japan	834 A.D. Danes raid England
809 A.D. Tibetan and Turkish troops support rebels in Sogdia against Arabs	810–850 A.D. Devapāla maintains Pāla empire; establishes contact with Śailendra rulers in Indonesia	819 A.D. Chinese Confucian master Han-yü composes memorial against Buddhism	840 A.D. Confederation of Slavs in E. Europe
810 A.D. T'ang Emperor Hsien-tsung requests return of Chinese territories	c. 820 A.D. Death of Śaṁkara Ācārya, Vedānta philosopher	820–824 A.D. Mu-tsung rules T'ang China	841 A.D. Monarchy established in Norway
814–836 A.D. Ral-pa-can rules, supports Dharma	840–885 A.D. Mihira Bhoja, Pratihāra ruler of Kanauj, pushes out Pālas	838 A.D. Last Japanese envoy to T'ang China	843 A.D. Treaty of Verdun divides Frankish kingdom
Standardization of translation terminology	846 A.D. Ceylon capital moved from Anurādhapura to Polonaruwa to defend against Chola invasions	841–846 A.D. Wu-tsung rules T'ang China; persecutes Buddhists, Nestorians, Mazdeans, Manicheans	844 A.D. Kenneth, King of Scots, defeats Picts
822 A.D. Treaty with China sworn at Ch'ang-an and lHa-sa; Tibet and China to retain own territories, cease hostilities		853 A.D. 1st book printed in China: Buddhist text	845 A.D. Vikings raid N. Germany
822 A.D. Treaty with Uighurs and 'Jang		858 A.D. Fujiwara clan takes power in Japan	867 A.D. Death of Pope Nicholas I; beginning of decline of papal authority
			867 A.D. Schism between Greek and Roman Churches

CHAPTER TWENTY

THE END OF THE EMPIRE

In the middle of the ninth century, Tibet was politically fragmented and lost the empire established by the great Dharma Kings.The sBa-bzhed explains that during the reign of Ral-pa-can certain nobles and Bon-pos had grown uneasy at the increasing influence Buddhism was exerting on the Tibetan culture. To discredit Buddhist practitioners and to isolate the king, the disgruntled faction forced Ral-pa-can's brother gTsang-ma, who had become a monk, into exile in Bhutan. Rumors were spread about Ral-pa-can's queen Nang-tshul and the monk Yon-tan-dpal, who had helped arrange the 822 treaty between China and Tibet and was an influential minister in the government. The gossip resulted in the assassination of Yon-tan and the queen's subsequent suicide.

Ral-pa-can's death is sometimes described as accidental — a fatal slip on the steps of the temple at Mal-dro, which cost him his life. The T'ang annals note that he died of an illness, but other sources say that Ral-pa-can was murdered by the ministers dBas rGyal-tho-re and Cog-ro Legs-smra.

In any case, Ral-pa-can was buried near Yar-lung valley, his tomb decorated with a remarkable stone lion carved in a style said by some modern scholars to be Persian. Ral-pa-can's brother,

Glang-dar-ma, was put on the throne with the backing of the Bon-pos and the rebellious clans sometime between 836 and 841, and sBas rGyal-tho-re became chief minister.

Glang-dar-ma's reign lasted only until 841 or 842, according to most sources. Towards the end, a persecution of Buddhism began that closed all official Buddhist institutions in central Tibet, and forced monks to return to lay life. Some Buddhist masters were killed, and monks were compelled to become hunters or butchers or in other ways violate the nonviolent code of the Buddhist tradition. The translation center where the Tibetan lo-tsā-bas had worked with the Indian paṇḍitas was destroyed, and temple doors were barred or plastered over. bSam-yas, Ra-mo-che, and Ra-sa gTsug-lag-khang were closed. The famous statue of Śākyamuni Buddha, brought to Tibet by Srong-btsan-sgam-po's Chinese wife nearly two hundred years before, was hauled out of the temple at lHa-sa and once again buried, while Kong-jo herself was portrayed as a demoness.

As the terror and destruction continued, a yogin named dPal-gyi-rdo-rje decided to end the reign of Glang-dar-ma, who, according to some accounts, had gone completely mad. Wearing a black cape with a white lining, and riding a white horse blackened with charcoal, dPal-gyi-rdo-rje approached the king at a ceremony and shot him down. He then headed east, running his horse through a river to restore its white color, and reversing his cape. Undetected, he escaped into the mountains of Khams.

THE HEIRS OF GLANG-DAR-MA

The confused aftermath of this assassination left central Tibet in complete disorder. Two young children, said to be descendants of the slain usurper Glang-dar-ma, were proposed as the rightful heirs. According to some sources, the eldest wife of Glang-dar-ma only claimed to be pregnant and found a little child whom she put forth as the true heir; he became known as Yum-brtan, "Depending on the Mother." The younger queen had a son she watched over carefully day and night with a light for fear that he would be harmed, and this child became known as 'Od-srung, "Guarded by Light."

The supporters of 'Od-srung gained the upper hand, and he was put on the throne while Yum-brtan and his descendants seem to have gained control over some regions in the northeastern part of central Tibet. 'Od-srung ruled from 843 to 905, according to Sa-skya historians, but the great empire built by Srong-btsan-sgam-po and Khri-srong-lde-btsan was hopelessly fragmented. It is not clear how far 'Od-srung's power extended, but he certainly had little influence over the vast reaches of the frontier lands. In the far northeast, Tibetan generals loyal to either 'Od-srung or Yum-brtan fought one another for years.

One general, known in the T'ang accounts as Chang Pi-pi, supported the new king 'Od-srung. He is described as a man from the Yang-t'ung tribes, a cultured intellectual, who, though he preferred literature to politics, had been made the governor of the Shan-chou region around Zi-ling in A-mdo. His opposition, Zhang Khon-bzher, considered 'Od-srung illegitimate, and the two fought each other until Chang Pi-pi was killed. But the victor was later attacked and destroyed by the Uighurs in 866.

LOSS OF THE EMPIRE

The Uighur Turks had begun spreading into the eastern edge of the Tarim basin after their own empire to the north had been destroyed in 840 by other Turkish tribes. The T'ang annals show that they took Tibetan forts in the east by 866, and by 872 controlled territories northeast of A-mdo. The descendants of the Uighurs still live north of Koko Nor, and are known as the Sha-ra-yu-gur.

Because Tun-huang remained under Tibetan control until after 842, its monasteries and clergy were spared the drastic persecution of Buddhism which began in China in that year. But between 848 and 861 Tun-huang revolted against Tibetan rule.

The eastern end of the Silk Route, one of the prized possessions of the Tibetan Empire, was lost.

The western end of the Silk Route also passed gradually out of Tibetan control and into the hands of the Qarluq Turks, but some Tibetan camps and forts may have remained as far west as Khotan until about 950. The tenth century Persian geography, the *Ḥudūd al-'Ālam,* indicates that some western regions were considered Tibetan territories even in that time, for Tibet was still said to control the roads from Kashgar to Khotan. Historical notices on Khotan in Chinese sources mention a Chinese mission to Khotan in 938 that found Tibetan camps all along the southern branch of the Silk Route up to Khotan, where Tibetan and Khotanese soldiers were still struggling — a century after the death of Ral-pa-can.

Though the power of the empire may have continued in outlying regions, formal alliances made with foreign powers lapsed, for no central government spoke for Tibet. Relations with China were disrupted, with only local chiefs from the northeast staying in contact with the T'ang government, which itself collapsed in 907. The various districts within Tibet ruled their own lands, while developing power structures around the most prominent clans and forging their own policies with their neighbors. Fragmentary records exist; for example, a letter found at Tun-huang from Kha-gan Teng-re-gye-pur, a Uighur ruler at Kan-chou, to the chief of the Tibetan district of rMa-khrom makes inquiries about previous correspondence from rMa-khrom. But it seems that records were not systematically kept by central authorities, or else have been lost. Thus our knowledge of this period is scanty. Perhaps local chronicles and family histories can eventually be pieced together to clarify this era.

DESCENDANTS OF 'OD-SRUNG AND YUM-BRTAN

We do know that the political situation in central Tibet remained unstable for some years. Disorders racked the center of the kingdom in the time of Glang-dar-ma's heir, 'Od-srung, and the tombs of the kings were vandalized about 877. 'Od-srung himself had his tomb built in Yar-lung, but he was the last Tibetan king to have done so. Though a great deal of damage had been done to the monastic institutions of Buddhism in Tibet, the official persecution came to an end with Glang-dar-ma. 'Od-srung was not hostile to Buddhism and is even reported by the Blue Annals to have built temples. His son dPal-'khor, who ruled from 906 to 924, also supported Dharma activity.

Old records show that while 'Od-srung ruled in gTsang, Yum-brtan controlled dBu-ru, and his descendants settled toward the east. Historians have preserved only lists of small princes and places they ruled; there is little information about most of these rulers. But these lists do reveal the process of fragmentation as the sons of one prince divided up his lands, or the sons of another prince moved off to establish three or four new estates.

Yum-brtan's son was Khri-lde-mgon, whose son mGon-brten had two sons, Rig-pa-mgon and Nyi-'od-dpal-mgon. According to the Deb-dmar-gsar-ma, their families spread to Rlung-shod, 'Phan-yul, and mDo-khams. Rig-pa-mgon had two sons, lDe-po and rDo-rje-'bar. lDe-po's descendants settled in Yar-stod and Thang-lha-brag. rDo-rje-'bar's son was dBang-phyug-btsan whose son Tshwa-na Ye-shes-rgyal-mtshan is known to have supported the Dharma. So did his son mNga'-bdag Khri-pa, whose own sons were A-tsa-rya, lHa dGe-slong, and lHa-btsun Bo-dhe-rā-dza, who met Atīśa at bSam-yas in the mid-eleventh century. A-tsa-rya's descendants became the chiefs at 'Phran-po, Grib-pa, sNye-thang, and Lum-pa. Bo-dhe-rā-dza's descendants included the chiefs of lHa 'Bri-sgang, as well as lHa-btsun Zhi-ba'i-'od and lHa-btsun sNgon-mo of bSam-yas. From them arose the family of Bu-tshal, the lHa-gling-dkar-pa, the princes of Tsha-rong and of lHa-rgya-ri. The princes of 'On were in lHa dGe-slong's line.

A different genealogy is recorded by Bu-ston, who gives Rig-pa-mgon's son as Khri-lde-po and his son as Khri-'od-po. Khri-'od-po's sons were A-tsa-rya, Gong-po-brtsan, and Gong-po-brtsegs. Rig-pa-mgon's brother, Nyi-'od-dpal-mgon, had a son mGon-spyod, and his son is Tsha-nal Ye-shes-rgyal-mtshan.

The lineage from 'Od-srung continued through dPal-'khor's sons and grandsons, who divided up what remained of their kingdom after a power struggle in 929. One son, bKra-shis-brtsegs, stayed in gTsang, and his sons founded what are called the "Lower Kingdoms." According to the Deb-dmar-gsar-ma, the descendants of the first son dPal-lde settled in lower mNga'-ris, and their families included the princes of rDzong-kha. The third and youngest son, sKyid-lde, was prince of rTa-nag in Shangs, and his six sons gave rise to the lords of Mus, 'Jad, and Nyang-stod in gTsang. The middle son 'Od-lde had four sons, the eldest of whom gave rise to the lords of gYag-sde and sTag-tshal in Nyang-stod. The descendants of his second son Khri-lde became the rulers of Shar-tsong-kha in mDo-smad. Leaders in the lHa Dong lineage are also connected to Khri-lde's line according to this source. The two youngest sons of 'Od-lde were Nyag-lde, who stayed in gYas-ru and sKyin-mkhar, and Khri-chung, who went to Yar-lung and 'Phying-ba.

Khri-chung's son was 'Od-skyid-'bar, who in turn had seven sons. Their descendants became the lords of gYu-thog, sNa-mo, 'Phying-ba, Don-mkhar, Thang-'khor, Khra-'brug, Bying, Byar, and Chu-mig-gog-po. One son took Bya-sa, and his son Jo-dga' had three sons, the oldest being the lHa-chen of Bya-sa, who built the temple of Bya-sa. The second son was Khri-dar-ma, who in turn had four sons, gTsug-lde, Khri-gtsug, Jo-bo rNal-'byor, and Jo-bo sMon-lam.

Jo-bo rNal-'byor took 'Ban-tshigs, while Jo-bo sMon-lam built sPu-gur-rdo. Khri-gtsug took dPe-sngon and Bya-sa, and his descendants were the rulers of Yar-mda'. Jo-bo rNal-'byor's son lHa-chen ruled at 'Chad-spyil in lHa-lung, while his other son Jo-'bag had in turn five sons. Of these, lHa-'gro-mgon became the head of 'Chad-spyil while Jo-bo Śākya-mgon founded gNas-chung. His son Jo-bo Śākya bKra-shis founded Pho-brang-rnying-ma and had two sons, lHa-brag-kha-pa and mNga'-bdag Grags-pa-rin-chen. This last one had three sons, Blo-gros-'od, lHa-btsun Tshul-khrims-bzang-po, and Jo-rgyal, whose son was Jo-'bar. Jo-'bar's sons were lHa-zur-khang-pa and lHa-khrom-po-pa of Rong-khrom-po. This line existed in Byar and in Sa-smad-spyan-rtsigs.

DESCENDANTS OF THE DHARMA KINGS IN THE WEST

When the kingdom was divided in 929, dPal-'khor's other son, Khri-lde Nyi-ma-mgon, went to the far west where he built the castle of Nyi-bzung in sPu-rang. According to the La-dwags rgyal-rabs, Nyi-ma-mgon married a woman of the 'Bro clan. Their sons founded kingdoms known as the "Upper Kingdoms": Rig-pa-mgon in Mar-yul, bKra-shis-mgon in sPu-rangs, and lDe-tsug-mgon in Gu-ge.

lDe-gtsug's sons were 'Khor-lde, who became the famous monk Ye-shes-'od, and Srong-lde, who had a son lHa-lde. lHa-lde had three sons, 'Od-lde, Zhi-ba'i-'od, and Byang-chub-'od. 'Od-lde's son rTse-lde had a son 'Bar-lde and from this line descended the Gu-ge rulers: bKra-shis-lde, Bhā-lde, and Nāga-lde, also known as Nāgadeva. Down to this king, according to the Deb-ther-dmar-po, one king ruled Gu-ge, sPu-rang, and Mar-yul.

Nāgadeva's son bTsan-phyug-lde became lord of Ya-tshe, where his lineage continued: bKra-shis-lde, Grags-btsan-lde, Grags-pa-lde, A-shog-lde, A-nan-smal, Re'u-smal, Ji-'dar-rmal, A-ji-rmal, Ka-lan-rmal, and Par-ti-rmal. In the fourteenth century, this lineage came to an end in Ya-tshe. But one bSod-nams-lde from sPu-rangs was then invited to be ruler. From the lineage of his son sPri-ti-rmal are descended rulers in Mang-yul, Nubra, Glo-pa, La-dwags, and Zangs-dkar, according to the records in the Deb-dmar-gsar-ma.

Rig-pa-mgon, also known as dPal-gyi-lde, founded his kingdom in Mar-yul. The La-dwags-rgyal-rabs traces this lineage through his son 'Gro-mgon and his descendants: lHa-chen Grags-pa-lde, lHa-chen Byang-chub-sems-dpa', lHa-chen rGyal-po, lHa-chen Utpala, lHa-chen Nag-lug, lHa-chen dGe-bhe, lHa-chen Jo-ldor, bKra-shis-mgon, lHa-rgyal, lHa-chen Jo-dpal, and lHa-chen dNgos-grub, who lived in the twelfth and thirteenth centuries.

These western kings made great contributions to the development of Buddhist culture in Tibet, but they did not have the political influence to reunite the whole kingdom. Neither did the rulers of eastern regions such as sDe-dge, Gling, 'Gu-log, and Mi-nyag, though they too made contributions to the Dharma, and even wielded considerable political power locally.

EASTERN KINGDOMS

The histories of the northern and eastern parts of Tibet after the fall of the empire also await a thorough investigation. A few important leaders and events are, however, discussed in a number of chronicles. In the ninth century, a descendant of the mGar clan, A-mnyes Byams-pa'i-dpal, settled in the east in the region of Gling, and from his line eventually emerged the rulers of sDe-dge. Twenty-three generations after A-mnyes Byams-pa'i-dpal lived 'Gar-chen Chos-sdings-pa, a contemporary of Sa-skya Paṇḍita (1182–1251). His descendant bSod-nams-rin-chen studied with Chos-rgyal 'Phags-pa. The lineage continued through rNgu-pa-sgu-ru, sTong-dpon Zla-ba-bzang-po, rGyal-ba-bzang-po, dKar-chen Byang-chub-'bum, bDe-chen bSod-nams-bzang-po and his brother rNgu-pa Chos-kyi-rdo-rje. One of bDe-chen's sons was Bo-thar, who in the fifteenth century sponsored the establishment of the monastery of lHun-grub-steng at sDe-dge. By this time sDe-dge expanded to include many lands that had belonged to the old region of Gling.

The history of Gling between the ninth and twelfth centuries is still obscure. Sometime toward the end of that period the famous ruler Ge-sar of Gling controlled a large region in the east. There are a number of conflicting dates for Ge-sar, which need to be carefully studied and compared. Some old records indicate that Ge-sar was contemporary with the twelfth century 'Bru lHa-rgyal, who had been residing in Gling until he feuded with Ge-sar. 'Bru lHa-rgyal then left Gling for 'Gu-log just to the north, and his descendants, A-'bum, 'Bum-g·yag, and Phag-thar (see chapter 8), established themselves there.

Farther north beyond 'Gu-log lay Hsi-hsia, known to Tibet as Mi-nyag. This kingdom, established in the eleventh century by the Mi-nyag King Se-hu, lasted some 260 years until Genghis Khan conquered Hsi-hsia in 1226. Between 'Gu-log and Mi-nyag lay Tsong-kha, ruled by a prince from the west, one rGyal-sras (997–1065). Sung annals note that he was said to belong to the ancient lineage of Tibetan kings, and he might be in the line of Khri-lde, a descendant of dPal-'khor and 'Od-srung who is reported to have founded kingdoms in Shar-tsong in mDo-smad.

The social and political structures in Tibet at this time are not well known. To determine how leadership changed over time and what kind of governments replaced the rule of the Dharma Kings will require more research. But clearly, despite the fragmented political situation and the loss of the empire, the ancient culture and the newly established Buddhist tradition both survived.

AN ERA OF FRAGMENTATION

This era of fragmentation in Tibet had parallels in numerous regions around the world. The T'ang dynasty, founded in the early seventh century at the same time that King Srong-btsan-sgam-po founded the Tibetan Empire, began to decline about 850 and collapsed in 907, to be followed by three centuries of disunity. The Uighur Orkhon Empire was destroyed in 840, and by the tenth century the Muslim Caliphs began to lose their hold on their empire as well. In Europe the empire of Charlemagne was divided amongst his heirs by the Treaty of Verdun, giving rise to the new kingdoms of Italy, Germany, and France.

In the mid-ninth century, a remarkable upsurge of religious persecution occurred all over Eurasia. The Byzantine Empire had seen the growth of monasticism in the eighth century, as well as a great controversy over the use of sacred images, and Church and State were often at odds. The works of the Christian Iconoclasts were destroyed under Empress Theodora shortly after the death of Emperor Theophilus in 842. In the Muslim empire, heretic Muslims, Jews, and Christians were being persecuted after Caliph al-Mu'tasim died in 842. And in China Emperor Wu-tsung began persecuting Buddhists in 842, soon extending the persecution to include Manicheans, Nestorian Christians, and Mazdeans.

THE DHARMA CONTINUES

Although central Tibet was hard hit by the official persecution under Glang-dar-ma, Buddhists continued as laymen, quietly studying and practicing in out-of-the-way hermitages. Many texts were preserved in hiding. Laymen who wanted to practice the Dharma were protected by powerful siddhas such as gNubs Sangs-rgyas-ye-shes and gNyags Jñānakumāra, who transmitted the Inner Tantra teachings to their disciples. This lineage was maintained by the Vidyādhara Zur-chen (b. 954), Paṇ-chen Rong-zom Chos-kyi-bzang-po (1012–1088), Zur-chung Shes-rab-grags-pa (1014–1074), and sGro-sbug-pa (1074–1134).

Other teachers and practitioners fled to the east where the persecution was not felt strongly. Old Buddhist centers there, such as Klong-thang in lDan-khog, established in the reign of Srong-btsan-sgam-po, became especially important centers where teaching continued uninterrupted and more openly. Vairotsana, the great Vajrayāna teacher from Khri-srong-lde-btsan's time, had spent many years in the east, and his disciples continued to transmit these essential teachings throughout the difficult period.

During the persecution, it was essential to preserve the texts of the Vinaya tradition, which was the foundation for the establishment of the formal Buddhist Sangha, and to ensure the continuation of the ancient ordination lineages. The Blue Annals explains how three monks from the monastery of dPal Chu-bo-ri accomplished this task: Rab-gsal of gTsang, gYo dGe-'byung of Pho-thong-pa, and sMar Śākyamuni of sTod-lungs, all of whom had been ordained by Cog-ro Klu'i-rgyal-mtshan.

These three gathered the crucial Vinaya and Abhidharma texts together and traveled west and north into Central Asia. These regions had been Tibetan territories for many decades and a stronghold of Buddhism for centuries. Old Buddhist centers like Khotan and Tun-huang were still active. The Uighur Turks were moving into the eastern regions beginning in about 840, and Buddhism was growing more and more popular among them as well. gYo, Rab, and sMar first stayed in Uighur territory, but they decided to travel east and south where the language and customs were not so foreign.

The three must have traveled through Mi-nyag, another old Tibetan territory where Buddhism was practiced, for they arrived in A-mdo in the northeastern corner of Tibet around Koko Nor lake. There they met a devoted young Buddhist practitioner who wished to become a monk. According to the Blue Annals, he was a reincarnation of Khri-sum-rje, the famous minister of the Dharma Kings. He asked the three Vinaya holders from central Tibet to ordain him, which they did, after locating the two additional monks necessary for the ceremony. Taking the religious name of dGongs-pa-rab-gsal, this new monk traveled north to study in Mi-nyag at Kan-chou and then south into other parts of Khams where he studied for many years. The Blue Annals gives many more details on how this monk became a great teacher.

Eventually, dGongs-pa-rab-gsal established a teaching center at Dan-tig mountain southeast of Koko Nor, where gYo, Rab, and dMar stayed as well. The Deb-ther-dkar-po notes that a pillar north of Zi-ling (Xining) is inscribed with the names of these three famous monks, who are buried in a temple at Zi-ling.

MONKS RETURN TO CENTRAL TIBET

The renown of dGongs-pa-rab-gsal, who later became known as Bla-chen, the Great Lama, spread even into western and central Tibet. Upon hearing that the Vinaya lineage continued unbroken in the east, young men traveled to A-mdo to be ordained by Bla-chen and his disciples.

Led by Klu-mes Tshul-khrims, this group of newly ordained monks returned from A-mdo to central Tibet to restore the Sangha. First they went to lHa-sa, which had been a great center of learning in the old days of the empire. But it had not yet recovered from the era of destruction, so they proceeded to bSam-yas and began to spread out from there. Several different princes in dBus are reported to have helped them, including Tshwa-na Ye-shes-rgyal-mtshan and mNga'-bdag Khri-pa, both descendants of Yum-brtan.

The monks refurbished old centers such as sKa-tshal and Yer-pa, and founded new monasteries such as gZhu-kun-dga'-ra-ba, Sol-nag Thang-po-che, rGyal-lug-lhas, and Tshong-'dus-tshogs-pa. The renewed freedom to practice openly and to ordain monks inspired the building of monastic centers where people of all walks of life could come to study and practice and participate in religious events.

The disappearance of the ordination practices from central Tibet lasted about seventy years according to most historians, though different lengths of time are calculated in various ways. The Blue Annals indicates that Bla-chen was born in a water-rat year of 832 and lived seventy-five years until the wood-pig year of 915. Sog-bzlog-pa's history says he was born in 855 and died in the earth-pig year of 939. He must have been very old when he ordained Klu-mes, who is said by the Blue Annals to have returned to central Tibet in the earth-tiger year of 978, sixty-four years before the arrival of Atīśa in 1042. Bu-ston mentions the possibility that Bla-chen actually ordained Grum Ye-shes-rgyal-mtshan, a contemporary of the last T'ang emperor Ai-ti (904–907), who in turn ordained Klu-mes.

BUDDHIST MASTERS ARRIVE FROM INDIA

About the time the ordination lineage was restored in central Tibet, the great Buddhist master Smṛtijñānakīrti together with Sūkṣmadīrgha arrived in Tibet. Their translator having died in Nepal, Smṛti remained in rTa-nag in gTsang for years as an unknown shepherd until he was recognized and invited to teach. Later he established a school in Khams at lDan-khog famous for Abhidharma and Tantra teachings, where he translated and taught many subjects.

Devoted Dharma practitioners in central Tibet also began searching out Buddhist masters. 'Brog-mi (d. 1074) spent thirteen years with great teachers and siddhas, finally returning to Tibet to teach and translate. His disciples established the Sa-skya-pa lineages. Mar-pa (1012–1097) visited India three times to study with Buddhist masters, and his disciples founded the bKa'-brgyud lineages, including the Karma, Tshal-pa, Phag-mo-gru, sTag-lung, 'Bri-gung, and others. Numerous Tibetans journeyed to Kashmir, Nepal, and India, and invited paṇḍitas to return with them to teach in Tibet.

Meanwhile, in the kingdoms of the far west established by the descendants of 'Od-srung, Buddhist teachings found support with dPal-'khor's grandson 'Khor-lde, who became a monk along with his two sons. Taking the name Ye-shes-'od, he handed the government of Gu-ge over to his brother Srong-lde. Translators and teachers were invited from Kashmir and India, and young men were sent to Kashmir to study. Because of the great difference in climate, most of the young Tibetans died. But seven years later, two returned. One of them, Rin-chen-bzang-po (958–1055), was soon to be known as the Great Translator. He trained new translators, who worked with the paṇḍitas invited by Ye-shes-'od. Temples and stupas were erected in Gu-ge and sPu-rang, and artisans invited from northwestern India and Kashmir. mTho-lding was built, as well as Ta-bo, A-lchi, and lHa-lung.

Toward the end of his life, Ye-shes-'od was captured by the Qarluq Turks, who by the mid-tenth century had become Muslims. Ye-shes-'od was offered his freedom for his weight in gold. But when the gold had been collected by his relatives, he insisted it be used instead as an offering to invite Buddhist masters to Tibet. His family followed his request and used the gold to establish the Dharma. His nephew lHa-lde and his great-nephews Byang-chub-'od, Zhi-ba'i-'od, and 'Od-lde continued to invite Buddhist teachers to Tibet.

Ye-shes-'od and the monk Byang-chub-'od had extended numerous invitations to one of the most renowned Buddhist masters of the day, Dīpaṁkaraśrījñāna, also known as Atīśa (982–1054). He was the son of the king of Sa-hor and the abbot of

Vikramaśīla university. During the reign of King 'Od-lde, in 1038, Byang-chub-'od sent men to invite the great master once again, and this time he consented. In 1042 he arrived in Gu-ge. He taught extensively in Tibet until his death in 1054, and his disciples, the most prominent among them being 'Brom-ston (1004–1064), established a Buddhist school known as the bKa'-gdams-pa.

These schools established during the eleventh and twelfth centuries as new teachings came into Tibet became known as the New Schools, or gSar-ma. The older tradition from the days of the great Dharma Kings now became known as the teachings of the Ancient Ones or rNying-ma. The establishment of all these schools and their monasteries and the rise of powerful patrons and influential families over the four centuries following the fall of the empire is a subject worthy of a book in itself.

After the era of the Dharma Kings, the predominant cultural activity for several centuries revolved around building monasteries, translation of sacred texts, and study with great teachers, such as the Kashmīri scholar Śākyaśrī (1127–1225), who arrived in Tibet in the thirteenth century. Thus political fragmentation did not destroy the cultural and religious efforts of the great kings, masters, and translators who had planted the seed of the Dharma in Tibet.

CONTRIBUTIONS OF ANCIENT TIBET

During the time of Srong-btsan-sgam-po, Mang-srong-mang-btsan, 'Dus-srong Mang-po-rje, Mes-ag-tshoms, Khri-srong-lde-btsan, Sad-na-legs, and Ral-pa-can, the Tibetan Empire had been won and lost. More long lasting than political power was the tremendous cultural advance stimulated by these kings. Tibetan Buddhist civilization brought benefits to the Tibetan people for centuries and preserved a precious heritage for the rest of the world. Within a few centuries after the Tibetan kings began to establish the Dharma in their land, Buddhism had been suppressed in China and destroyed in India and Central Asia. By the eleventh century, Muslim Qarakhanid Turks held Khotan, and Muslim Ghaznavid Turks had conquered the Punjab. In the thirteenth century, a Muslim Turko-Afghan dynasty established the Delhi Sultanate in India. But the efforts of the Dharma Kings had assured that the Buddhist teachings were already successfully transplanted to Tibet.

The accomplishments of the Tibetan Dharma Kings are difficult to measure. The modern world is only just beginning to realize the value of the knowledge that Tibet has preserved as East meets West. In the future, the contributions that these kings and the Tibetan people have made to world civilization may be more widely known.

ASIAN POWERS AFTER 900 A.D.

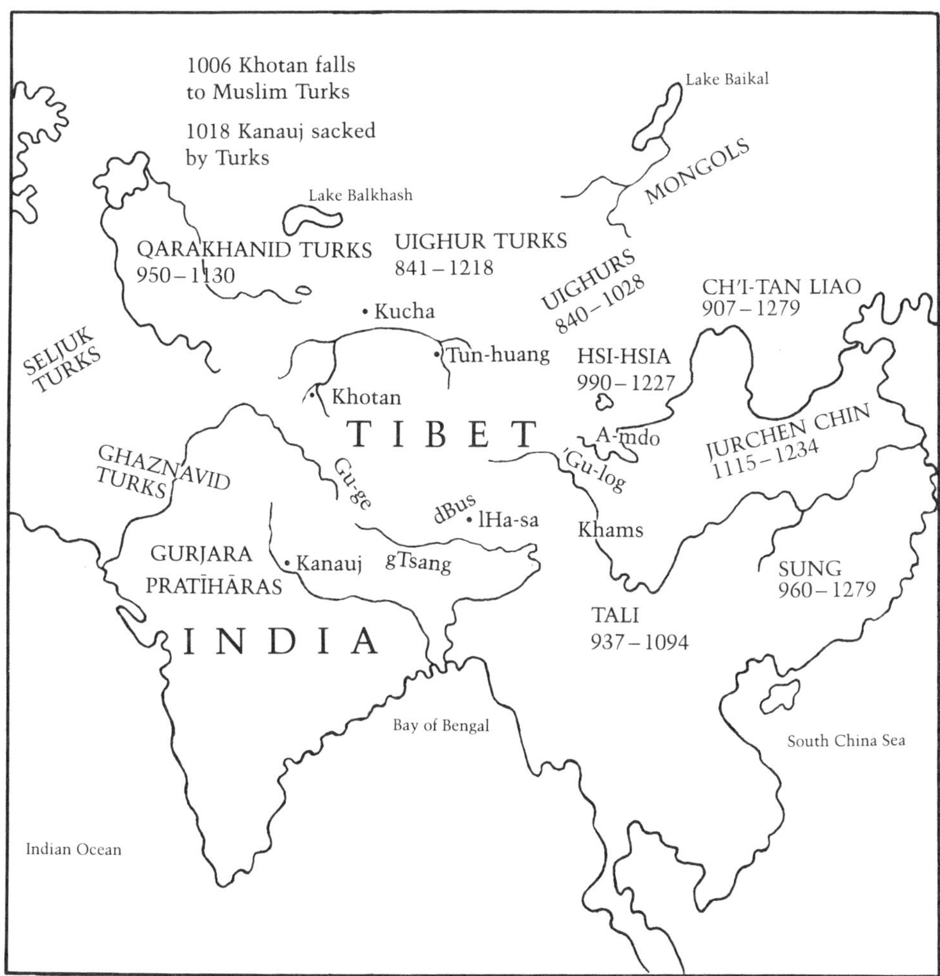

In the ninth century, the Tibetan Empire began to disintegrate after the assassinations of Ral-pa-can and Glang-dar-ma. Dar-ma's heirs fought over central Tibet, while control of the eastern and western territories of Tibet disappeared, and the administration of the vast frontier territories in Central Asia lapsed. As the Uighurs and other Turkish tribes rose to power, they took over the Central Asian cities that had belonged to Tibet. By the eleventh century, Muslim Turks held not only Central Asia, but also parts of far western India.

THE ROYAL TOMBS

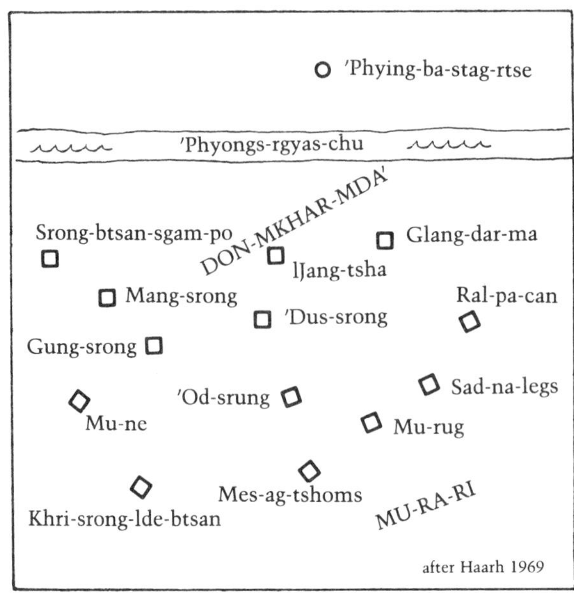

Royal tombs of eleven of the Tibetan btsan-pos and two princes since the time of Srong-btsan-sgam-po can be identified at 'Phyongs-rgyas east of 'Phying-ba stag-rtse. Across the river in the valley known as Don-mkhar-mda', which is close by the foothills of Mu-ra-ri, numerous large mounds can still be seen. Modern scholars have connected locations of the tombs with accounts in Tibetan histories describing the placement of the royal tombs, but differing opinions exist.

THE DHARMA SURVIVES

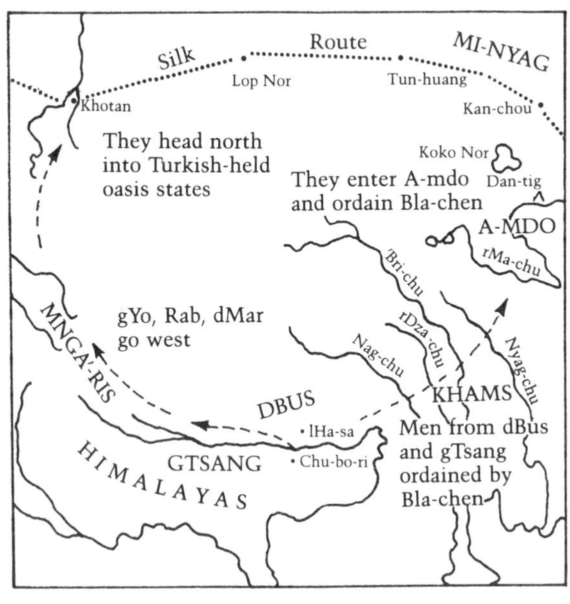

Although the Tibetan Empire collapsed in the mid-ninth century after the assassinations of two kings, and a persecution of Buddhists began in central Tibet, Buddhist teachings so painstakingly established by the determined Dharma Kings were not lost. In central Tibet, lay practitioners continued quietly under the protection of great masters, while the ordination lineage from Śāntaraksita was preserved in the east through the efforts of gYo, Rab, dMar, and Bla-chen.

DESCENDANTS OF THE DHARMA KINGS

according to the Deb-dmar-gsar-ma

'OD-SRUNG'S LINEAGE

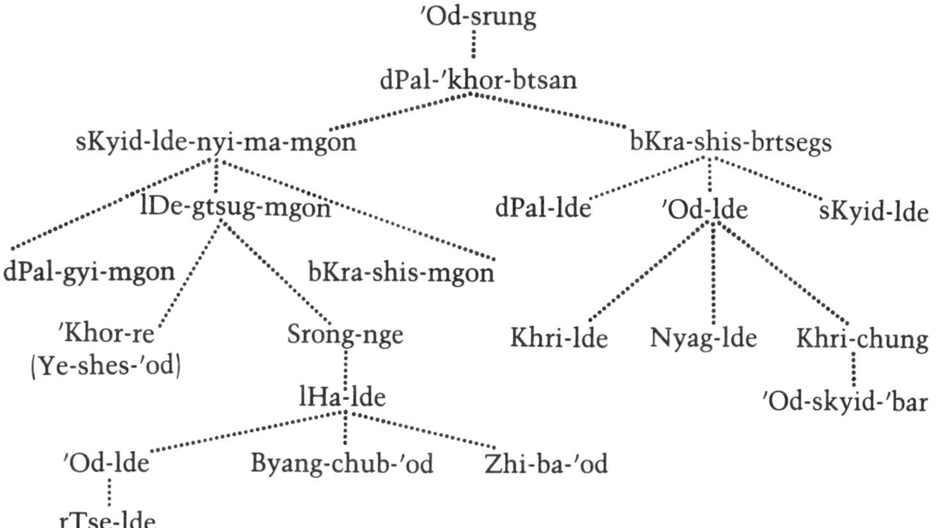

YUM-BRTAN'S LINEAGE

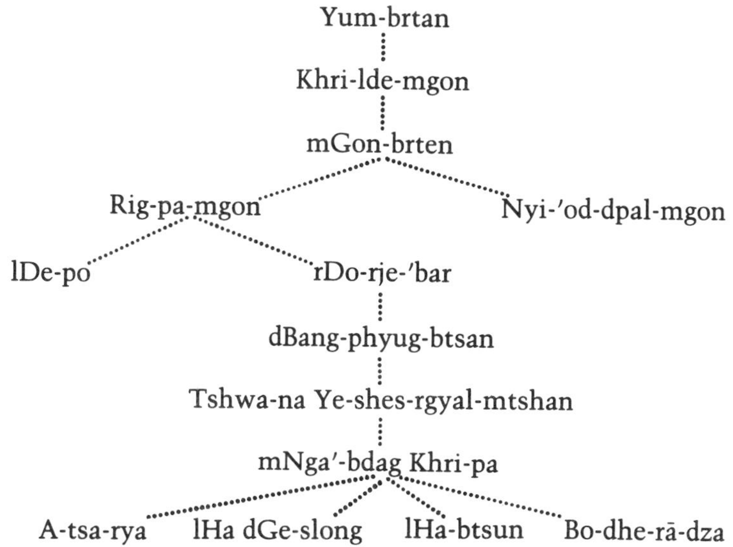

THE NINTH AND TENTH CENTURIES

TIBET	ASIA	FAR EAST	WEST
836 /838 A.D. Ral-pa-can assassinated Glang-dar-ma takes throne Persecution of Buddhists begins 841 A.D. Glang-dar-ma assassinated 843 – 905 A.D. 'Od-srung rules central Tibet 848 – 861 A.D. Tun-huang revolts 866 A.D. Uighurs take Tibetan forts in eastern Central Asia 877 A.D. Tombs of kings vandalized 906 – 924 A.D. dPal-'khor rules central Tibet 929 A.D. Rebellion in central Tibet: Nyi-ma-mgon goes west to mNga'-ris c. 975 A.D. 'Khor-re (Ye-shes 'Od) rules Gu-ge 1032 A.D. rGyal-sras rules A-mdo, Tsong-kha	9th century Qarluq Turks control N.W. Central Asia c. 820 A.D. Devapāla rules Bengal c. 840 – 885 A.D. Mihira Bhoga, Gurjara king, rules most of N. India 840 A.D. Uighur Empire destroyed by Kirghiz and Qarluq Turks; Uighurs into Tarim basin 841 A.D. Burmese establish capital at Pagan 841 – 1218 A.D. Uighurs control central Tarim basin oasis states 937 – 1094 A.D. Nan-chao Ta-li kingdom rules Yunnan 990 – 1227 A.D. Hsi-hsia nation rules eastern Central Asia; borders China and Tibet 1018 A.D. Turkish Muslim Mahmud of Ghazni sacks Kanauj in India 1044 A.D. Anōratha unites Burma	820 – 824 A.D. Mu-tsung rules T'ang China 838 A.D. Japan discontinues embassies to China: end of three centuries of cultural borrowing 841 A.D. Persecution of Manicheans, Buddhists, Nestorians, and Mazdeans begins in China 866 – 1160 A.D. Fujiwara Period in Japan 907 – 959 A.D. Period of Five Dynasties follows collapse of T'ang dynasty in China 935 A.D. Kingdom of Koryŏ founded in Korea 937 – 1127 A.D. Ch'i-tan tribes establish Liao kingdom in Manchuria, occupy north China 960 – 1127 A.D. Sung dynasty in China: Golden Age of landscape painting; scholarship flourishes	862 A.D. Rurik, 1st Russian grand prince, founds Novgorod 843 A.D. Treaty of Verdun: kingdoms of France, Germany, and Italy arise c. 850 A.D. Collapse of Mayan civilization in Middle America; rise of Toltecs 878 A.D. Danes dominate N. Britain and Scotland; Alfred the Great rules the south 936 A.D. Power of Moslem Caliphs in Baghdad declines 925 – 961 A.D. Germany gains control of N. Italy and central Europe 969 – 1171 A.D. Islamic Fatimid dynasty rules Egypt 1055 A.D. Seljuk Turks capture Baghdad 1096 A.D. First Crusade; Franks invade Near East

APPENDICES

PLACE-NAMES

TIBETAN	PINYIN
Koko Nor	Qinghai
Zi-ling	Xining
Ra-rgya	Ragya
Co-ne	Jone
Bla-brang	Xiahe
mDzo-dge	Zoge
Tshwa'i-'dam	Qaidam
Khyer-dgun-mdo	Yushu
Nang-chen	Nangqen
gSer-thal	Serthar
lDan-khog	Danko
Sur-mang	Surmang
sDe-dge	Dege
rDzogs-chen	Zogqen
dPal-yul	Bayu
Kon-jo	Gonjo
Brag-g·yab	Zhag'yab
Ka-mdze	Garze
lTa'o	Dawu
rGyal-mo-rong	Jinchuan
Brag-mgo	Luhuo, Zhaggo
mGar-thar	Qianning
Nyag-rong	Xinlong
Ri-bo-che	Riwoqe
gTing-chen	Dengqen
Chab-mdo	Qabdo
Tsha-ba-mdzo-sgang	Zogang
dMar-khams	Markham
'Ba'-thang	Bathang
Li-thang	Lithang
Dar-rtse-mdo	Dardo, Kangding
Nag-chu-kha	Nagqu
Sog-rdzong	Sog Xian

PLACE-NAMES

TIBETAN	PINYIN
lHa-ri-mgo	Lhari
lHo-rong-rdzong	Lhorong
sPo-bo	Bomi
Kong-po	Gongbo
rGya-mdo	Gyamda
sNe'u-gdong	Nedong
Zangs-ri-khang-dmar	Sangri
'Phyongs-rgyas	Qonggya
Gong-dkar	Gonggar
Chu-shul	Quxur
sNye-mo	Nyemo
lHa-sa	Lhasa
Mal-dro	Mazho
Phu-mdo	Lhunzhub/Poindo
'Bri-gung	Zhigung
Yangs-pa-can	Yangbajan
sPa-rnam	Banam
Rin-spungs	Rinbung
rGyal-rtse	Gyaze
Khang-dmar	Kangmar
gZhis-ka-rtse	Xigaze
Sa-skya	Sagya
rNam-gling	Namling
lHa-rtse	Lhaze
Phun-tshogs-gling	Puncoling
gTing-skyes	Dinggye
Ding-ri	Dingri
Sa-dga'	Saga
sPu-rang	Busheng
sGar-dbyar-sa	Garyarsa
sGar-dgun-sa	Gar
bDe-mchog	Demqo
Ru-thog	Ruto

GEOGRAPHICAL NAMES

TIBETAN	OLD MAPS	PINYIN
Kun-lun	K'un-lun	Kunlun Shan
name uncertain	Altyn Tagh	Altun Shan
Mi-nyag	Nan-shan	Qilian Shan
Ko-ko-shi-li	Kokoshili	Hohxil Shan
Ba-yan-ha-ra	Bayankara	Bayanhar Shan
A-mnyes-rma-chen	Amnye Machin	Anye Maqen Shan
lDang-la	Tanglha	Tanggula Shan
sMar-khams-sgang	not noted	Ningjing Shan
Tsha-ba-sgang	not noted	Taniantaweng Shan
Mi-nyag-rab-sgang	Ta-hsueh-shan	Daxue Shan
gYar-mo-sgang	not noted	Shaluli Shan
Zab-mo-sgang	not noted	Chola Shan
gNyan-chen-thang-lha	Nyenchen Tangla	Nyainqen Tanglha Shan
Gangs-ri-ti-se	Kailash Range	Gangdise
mNga'-ris-gangs-ri	Alung Gangri	Nganglong Kangri
Klu-chu	Tao	Tao Ho,Lu Qu
rGyal-mo-dngul-chu	Ta-chin-ch'uan	Dajin, Dadu He
She-chu, sMe-chu	She-chu	Xianshui
rMa-chu	Ma-chu	Ma Qu
Dza-chu, Nyag-chu	Nya-chu	Yalong
'Bri-chu	Dre-chu	Zhi Qu, Tongtian
rDza-chu	Dza-chu	Za Qu
dNgul-chu, Nag-chu	Ngu-chu, Nak-chu	Nu Jiang
gTsang-po	Tsang-po	Zangpo
sKyid-chu	Kyi-chu	Gyi Qu
Seng-ge-kha-'bab	Senge, Indus	Shiquan
sKya-rengs-mtsho	Tsaring Lake	Gyaring Hu
sNgo-rengs-mtsho	Oring Lake	Ngoring Hu
mTsho-sngon	Koko Nor	Qinghai Hu
gNam-mtsho	Namtso, Tengri Nor	Nam Co
Yar-'brog-mtsho	Yamdrok Tso	Yamzho Yumco
Pang-gong-mtsho	Pangong Lake	Bangong Co

DATES FOR THE DHARMA KINGS

	BOD-KYI-RGYAL-RABS		DEB-DMAR		NGO-MTSHAR-RGYA-MTSHO	
Srong-btsan-sgam-po						
birth	fire-ox	557/617	fire-ox	557/617	fire-ox	557/617
reign	earth-ox	569/629	earth-ox	569/629		
death	earth-dog	638/698	earth-bird	649/709		
Kong-jo arrives						
Kong-jo dies						
Gung-srong						
birth						
reign	at 13		at 13			
death	at 18		at 18		at 18	
Mang-srong						
birth	fire-dog	626/686	fire-dog	626/686	fire-dog	626/686
reign	earth-dog	638/698	earth-dog	638/698	earth-dog	638/698
death	water-rat	652/712	water-rat	652/712	water-rat	652/712
'Dus-srong						
birth	water-rat	652/712	water-rat	652/712	water-rat	652/712
reign	water-rat	652/712				
death	iron-dragon	680/740	iron-dragon	680/740	iron-dragon	680/740
Mes-ag-tshoms						
birth	iron-dragon	680/740	iron-dragon	680/740	iron-dragon	680/740
reign	iron-dragon	680/740				
death	water-horse	742/802	water-horse	742/802		
Kim-sheng arrives						
Kim-sheng dies						
Khri-srong-lde-btsan						
birth	iron-horse	730/790	iron-horse	730/790	iron-horse	730/790
reign	water-horse	742/802	water-horse	742/802		
death	wood-ox	785/845	wood-ox	785/845	wood-ox	785/845
Mu-ne						
birth	water-tiger	762/822	water-tiger	762/822		
reign	fire-tiger	786/846			fire-tiger	786/846
death	fire-hare	787/847	fire-hare	787/847		
Ju-tse						
reign						
death						
Sad-na-legs						
birth	wood-dragon	764/824	wood-dragon	764/824	wood-dragon	764/824
reign	fire-hare	787/847	fire-hare	787/847	fire-hare	787/847
death	fire-bird	817/877	fire-bird	817/877	fire-bird	817/877
Ral-pa-can						
birth	fire-dog	806/866	fire-dog	806/866	fire-dog	806/866
reign	fire-bird	817/877	fire-bird	817/877	fire-bird	817/877
death	iron-bird	841/901	iron-bird	841/901	iron-bird	841/901
Glang-dar-ma						
birth	water-sheep	803/863				
reign	iron-bird	841/901				
death	water-dog	842/902				

DATES FOR THE DHARMA KINGS

	GSAL-BA'I-ME-LONG		BU-STON		BOD-KYI-LO-RGYUS	
Srong-btsan-sgam-po						
birth			fire-ox	557/617	fire-ox	617
reign					earth-ox	629
death			earth-dog	638/698	earth-dog	698
Kong-jo arrives					fire-monkey	636
Kong-jo dies						
Gung-srong						
birth					earth-snake	669
reign					iron-snake	681
death					fire-pig	687
Mang-srong						
birth	fire-dog	626/686			fire-dog	686
reign					earth-dog	698
death	water-rat	652/712			water-rat	712
'Dus-srong						
birth	water-rat	652/712			water-rat	712
reign					water-rat	712
death	iron-dragon	680/740			iron-dragon	740
Mes-ag-tshoms						
birth	iron-dragon	680/740			iron-dragon	740
reign					iron-dragon	740
death	water-horse	742/802				
Kim-sheng arrives						
Kim-sheng dies						
Khri-srong-lde-btsan						
birth					iron-horse	790
reign			at 13			
death			at 69		earth-tiger	858
Mu-ne						
birth	water-tiger	762/822	earth-rat	748/808		
reign			water-tiger	762/822	fire-rat	856
death	iron-horse	790/850	wood-dragon	764/824	fire-ox	857
Ju-tse						
reign						
death						
Sad-na-legs						
birth						
reign			at 4		fire-ox	857
death	fire-bird	817/877				
Ral-pa-can						
birth	fire-dog	806/866	fire-dog	806/866	fire-dog	866
reign			water-hare	823/883		
death	iron-bird	841/901	iron-bird	841/901		
Glang-dar-ma						
birth						
reign			iron-bird	841/901	iron-bird	901
death					fire-tiger	906

DATES FOR THE DHARMA KINGS

	DEB-SNGON		TUN-HUANG		DEB-DMAR RGYA-DEB	T'ANG
Srong-btsan-sgam-po						
birth	earth-ox	569				
reign	iron-ox	581				
death	iron-dog	650	dog	650	650	650
Kong-jo arrives	iron-ox	641	ox	641	641	641
Kong-jo dies	iron-dragon	680	sheep	683	680	680
Gung-srong						
birth						
reign						
death						
Mang-srong						
birth	earth-dog	638			638	
reign	iron-dog	650	dog	650	650	650
death	earth-hare	679	rat	676	679	679
'Dus-srong						
birth			rat	676		
reign	earth-hare	679	rat	676		679
death	wood-dragon	704	dragon	704	704	705
Mes-ag-tshoms						
birth			dragon	704	704	699
reign	wood-snake	705	dragon	704		705
death	wood-sheep	755	sheep	755	755	755
Kim-sheng arrives	water-rat	712	dog	710	710	710
Kim-sheng dies			hare	739	741	741
Khri-srong-lde-btsan						
birth			horse	742		
reign	wood-sheep	755	sheep	755	755	755
death	iron-monkey	780			780	780
Mu-ne						
birth			rat	760		
reign	iron-monkey	780			780	780
death	fire-ox	797			797	797
Ju-tse						
reign	fire-ox	797			797	797
death	wood- monkey	804			804	804
Sad-na-legs						
birth						
reign	wood- monkey	804			804	804
death	wood-horse	814			809	817
Ral-pa-can						
birth						
reign	wood-horse	814				817
death	fire-dragon	836			836	838
Glang-dar-ma						
birth						
reign	fire-dragon	836			836	838
death	iron-bird	841			841	842

GUIDE TO TOPICS, MAPS, CHARTS, AND TIMELINES

TOPICS

PART ONE: THE LAND

MAPS

THE LAND

THE PEOPLE

THE EMPIRE

CHARTS

TIMELINES

SOURCES FOR FURTHER STUDY

Highlights of Tibetan history from ancient times up to the ninth century can be found in old Tibetan materials from Central Asia, inscriptions, edicts, and documents from the era of the Dharma Kings, as well as in Tibetan histories. Some of these Tibetan accounts have been translated into Western languages. Additional information can be found in Chinese records from the Han, Sui, and T'ang dynasties, and in Khotan texts and Arab accounts. Several modern works on Tibetan history contain interesting material on ancient times, while Western scholarly publications offer detailed discussions of specific issues. Scientific information on the development of the plateau can be found in scientific journals and publications from various academies of science and U.S. governmental agencies, and in the Beijing press.

TIBETAN MATERIALS FROM TUN-HUANG AND CENTRAL ASIA

Choix de documents tibétains. Spanien, Ariane and Imaeda, Yoshiro. Paris, 1979.

Tun-huang annals, Tun-huang chronicles, Tun-huang rgyal-rabs in Documents de Touen-houang relatifs à l'histoire du Tibet. J. Bacot, F.W. Thomas, Ch. Toussaint. Annales du Musée Guimet. 51. 1940.

Documents on Sha-chu, Tshal-byi, Khotan, Dru-gu, and administrative and military records. Tibetan Literary Texts and Documents Concerning Chinese Turkestan. Parts II, III, IV. F.W. Thomas. Oriental Transl. Fund. N.S. 37, 40, 41. Royal Asiatic Society. 1951, 1955, 1963.

Tun-huang material on folklore. Folk-Literature from Northeastern Tibet. F.W. Thomas. Abhandlungen der Deutschen Akademie der Wissenschaften zu Berlin. Sprachen, Literatur und Kunst. Nr. 3. 1952.

Tun-huang text on folklore. Nam, An Ancient Language of the Sino-Tibetan Borderland. F.W. Thomas. Publ. of the Philological Soc. 14. 1948.

Tun-huang material published in Western journals. Transliterations, translations, or summaries:

Lalou, M. Journal Asiatique. 243. 1955.
Bacot, J. Journal Asiatique. 244. 1956.
Lalou, M. Journal Asiatique. 246. 1958.
Richardson, H.E. Bulletin of Tibetology. 2. 1965.
Richardson, H.E. Bulletin of Tibetology. 4. 1969.
Stein, R.A. Journal Asiatique. 258. 1970.
MacDonald, A. in Études tibétaines dédiées à Marcelle Lalou. 1971.
Stein, R.A. in Études tibétaines dédiées à Marcelle Lalou. 1971.
Richardson. H.E. In Buddhist Thought and Asian Civilization. 1977.

EDICTS AND INSCRIPTIONS FROM TIBET

rGyal-lha-khang rdo-rings. In H.E. Richardson. Journal of the Royal Asiatic Society. 1957.

sKar-chung rdo-rings. In H.E. Richardson. Journal of the Royal Asiatic Society of Bengal. 1949.

Khra-'brug dril-bu. In H.E. Richardson. Journal of the Royal Asiatic Society. 1953.

Kong-po rdo-rings. In H.E. Richardson. Journal of the Royal Asiatic Society. 1953.

'Phyongs-rgyas rdo-ring. In H.E. Richardson. Journal of the Royal Asiatic Society. 1964.

Sad-na-legs rdo-rings. In G. Tucci. Serie Orientale Roma. Is. M. E. O. I. 1950. Also in H.E. Richardson. Journal of the Royal Asiatic Society. 1969.

bSam-yas dril-bu and rdo-rings. In Tucci. 1950. Richardson. 1949.

gTsug-lag-khang rdo-rings. In H.E. Richardson. Journal of the Royal Asiatic Society. 1952 and 1978. Also in Fang-kuei Li. T'oung Pao. 44. 1956.

mTshur-phu rdo-rings. In Richardson. 1949.

Zhol rdo-rings. In H.E. Richardson. Prize Publication Fund. 19. Royal Asiatic Society of Great Britain and Ireland. 1952.

Zhwa'i-lha-khang rdo-rings. In H.E. Richardson. Journal of the Royal Asiatic Society. 1952, 1953.

First Edict of Khri-srong-lde-btsan. In mKhas-pa'i-dga'-ston. Translation in Tucci. 1950.

Second Edict of Khri-srong-lde-btsan. In mKhas-pa'i-dga'-ston. Translation in Tucci. 1950.

Edict of Sad-na-legs. In mKhas-pa'i-dga'-ston. Translation in Tucci. 1950.

TEXTS FROM THE ERA OF THE DHARMA KINGS

Ma-ṇi bka'-'bum. Srong-btsan-sgam-po. Discovered by Grub-thob dNgos-grub, mNga'-bdag Nyang, Śākya 'Od-zer twelfth century.

bKa'-thang-sde-lnga. Discovered by O-rgyan-gling-pa (1323–1360).

sBa-bzhed gtsang-ma and zhabs-brtag-ma. sBa gSal-snang. Ninth century with later additions.

bKa'-chems-ka-khol-ma. Last Testament of Srong-btsan-sgam-po. Discovered by Atīśa eleventh century.

Li'i-yul-lung-bstan-pa. bsTan-'gyur. Ny. 4202. Translation in Emmerick. 1967.

dGra-bcom-pa dGe-'dun-'phel-gyis lung-bstan-pa. bsTan-'gyur. Ny. 4201. Translation in Thomas. 1935.

Li-yul-chos-kyi-lo-rgyus. Tun-huang text published in Emmerick. 1967. and Thomas. 1935.

TIBETAN HISTORIES

Bod-kyi-rgyal-rabs. Grags-pa-rgyal-mtshan (1147–1216). Translation in Tucci. 1947.

Bod-kyi-rgyal-rabs. 'Gro-mgon Chos-rgyal 'Phags-pa (1235–1280). Translation in Tucci. 1947.

Bu-ston Chos-'byung. Bu-ston Rin-chen-grub (1290–1364). Translation in Obermiller. 1931, 1932.

Chos-'byung Rin-po-che'i-gter-mdzod thub-bstan-gsal-bar-byed-pa'i nyi-'od. Kun-mkhyen Klong-chen-pa (1308–1364).

Deb-ther-dmar-po. Tshal-pa Kun-dga'-rdo-rje (1346).

rGyal-rabs-rnams-kyi-'byung-tshul gsal-ba'i-me-long chos-'byung. bSod-nams-rgyal-mtshan (1312–1375).

Deb-ther-sngon-po. 'Gos Lo-tsā-ba gZhon-nu-dpal (1392–1481). Translation in Roerich. 1949.

Rlangs Po-ti-bse-ru. Fifteenth century.

Chos-'byung. Ratna-gling-pa (1403–1478).

Deb-ther-dmar-po-gsar-ma. bSod-nams-grags-pa (1478–1554).

Chos-'byung mKhas-pa'i-dga'-ston. dPa'-bo gTsug-lag-phreng-ba (1504–1566).

Chos-'byung bsTan-pa'i-padma-rgyas-pa'i-nyin-byed. Padma dkar-po (1527–1592).

rNam-thar-yid-kyi-mun-sel. Sog-zlog-pa Blo-gros-rgyal-mtshan (1552–1624).

rGya-gar-chos-'byung. Tāranātha (b. 1575).

Chos-'byung Ngo-mtshar-rgya-mtsho. sTag-lung-pa Ngag-dbang-rnam-rgyal (1571–1626).

Deb-ther rdzogs-ldan-gzhon-nu'i-dga'-ston. rGyal-ba lnga-pa-chen-po (1617–1682).

Bai-ḍūrya-dkar-po. sDe-srid Sangs-rgyas-rgya-mtsho (1653–1705).

Srid-pa'i-rgyud-kyi-kha-byang-chen-mo. Bon-po Chronicle.

La-dwags-rgyal-rabs. Seventeenth century.

sDe-dge-rgyal-rabs. Tshe-dbang rdo-rje-rig-'dzin (b. 1786).

Shes-bya-kun-khyab. 'Jam-mgon Kong- sprul Blo-gros-mtha'-yas (1813–1899).

'Gu-log Chos-'byung. Kyi-lhong-rnam-'gyur.
Nineteenth century.

dBus-gtsang-gi-gnas-rten-rags-rim-gyis mtshan-byang-mdor-bsdus Dad-pa'i-sa-bon. 'Jam-dbyangs-mkhyen-brtse'i-dbang-po (1820–1892). Translated in Ferrari. 1958.

Chos-'byung. Zhe-chen-rgyal-tshab (b. 1871).

Bod-kyi-lo-rgyus. 'Jigs-'bral-ye-shes-rdo-rje (1904–).

Deb-ther-dkar-po. dGe-'dun-chos-'phel (1905–1951). Translated in Choephel. 1978.

Bod-kyi-srid-don-rgyal-rabs. Zhwa-sgab-pa (1976). Translation in Shakabpa. 1967.

CHINESE MATERIALS IN TRANSLATION

Han, Sui, T'ang records on Central Asia. Summaries in Ancient Khotan. I. M.A. Stein. 1907.

Han dynasty records on Ch'iang tribes. Cited in Thomas. 1948. R.A. Stein. 1957–58. R.A. Stein. 1961.

Sui and T'ang annals, encyclopedia of 1013. Translations in Documents sur les Tou-kiue occidentaux. Édouard Chavannes. 1903.

T'ang annals on Tibet. Translation as Histoire Ancienne du Tibet. Paul Pelliot. 1961.

T'ang records on India and Nepal. Summarized by S. Lévi. Journal Asiatique. 15. 1900.

T'ang and later sources on Tibet. In Tibet. Geographical, Ethnological, Historical Sketch Derived from Chinese Sources. W. W. Rockhill. Extract from Journal of Royal Asiatic Society. 1891.

Records on Yunnan and Nan-chao. Summaries by Blackmore. Journal of Southeast Asian History. 1. 1960.

Records on Yunnan and Nan-chao. Summaries by E. Rocher. T'oung Pao. 10. 1899.

Si-yu-ki. Hsüan-tsang. Buddhist Records of the Western World. Translated by Samuel Beal. Two volumes. 1885.

Tzu-chih t'ung-chien. Ssu-ma Kuang (1085). Cited in Chavannes. 1903. and Beckwith. 1980.

Wei-tsang t'ung-chih. 1896. Cited in Li, Tieh-tseng. Tibet Today and Yesterday. 1960.

ARAB AND PERSIAN HISTORIES

Ḥudūd al-'Ālam. The Regions of the World. 982 A.D. Persian geography. Translated by Vladimir Minorsky. 1970.

al-Ya'qūbī. Les Pays. Translated by Gaston Wiet. Textes et traductions d'auteurs orientaux. Publications de l'Institut français d'archéologie orientale. 1937.

al-Ya'qūbī. Tārīkh al-Ya'qūbī. Cited in Beckwith. 1980. Petech. 1977.

TRANSLATIONS OF TIBETAN HISTORIES

Chattopadhyaya, Alaka and Lama Chimpa. Tāranātha's History of Buddhism in India. 1970.

Choephel, Gedun. The White Annals. Translated by Samten Norboo. 1978.

Emmerick, R.E. Tibetan Texts Concerning Khotan. 1967.

Ferrari, A. mK'yen brtse's Guide to the Holy Places of Central Tibet. Serie Orientale Roma. 16. 1958.

Francke, A.H. Antiquities of Indian Tibet. Part Two. Chronicles of Ladakh. Archeological Survey of India. New Imperial Series. 50. 1926.

Kolmaš, Josef. A Genealogy of the Kings of Derge. Tibetan text with introduction. Czechoslovak Academy of Sciences. Dissertationes Orientales. 12. 1969.

Obermiller, E. History of Buddhism by Bu-ston. Materialien zur Kunde des Buddhismus. 18, 19. 1931, 1932. Suzuki Research Foundation. Reprint Series 5.

Roerich, G.N. The Blue Annals. Two volumes. 1949.

Shakabpa, W.D. Tibet, A Political History. 1967.

Stein, R.A. Une chronique ancienne de bSam-yas: sBa-bzhed. Tibetan and French summary. Publications de l'Institut des hautes études chinoises. Paris. 1961.

Thomas, F.W. Tibetan Literary Texts and Documents Concerning Chinese Turkestan. Part I. Oriental Translation Fund. New Series. 32. Royal Asiatic Society. 1935.

Tucci, G. Deb-ther-dmar-po-gsar-ma. Serie Orientale Roma. 24. 1971.

Wylie, T.V. The Geography of Tibet According to the 'Dzam-gling-rgyas-bshad. Serie Orientale Roma. 25. 1962.

WESTERN RESEARCH ON ANCIENT TIBET

HISTORICAL STUDIES

PREHISTORY

Chatterji, S.K. Journal of the Royal Asiatic Society of Bengal. Letters. 16. 1950.

MacDonald, A. Journal Asiatique. 241. 1953.

Tucci, G. The Ancient Civilization of Transhimalaya. 1973.

Stein, R.A. Journal Asiatique. 251. 1963.

Tucci, G. East and West. New Series. 27. 1977.

ANCIENT TRIBES

Lalou, M. Journal Asiatique. 253. 1965.

Richardson, H.E. Bulletin of Tibetology. 4. 1967.

Richardson, H.E. Tibet Journal. 2. 1977.

Kania, I. Tibet Journal 3. 1978.

Richardson, H.E. Bulletin of Tibetology. 4. 1967.

Richardson, H.E. Tibet Journal. 2. 1977.

Stein, R.A. Annuaire 1957–58 de l'École pratique des hautes études. Section des sciences religieuses.

Stein, R.A. Les tribus anciennes des marches sino-tibétaines. Bibliothèque de l'Institut des hautes études chinoises. 15. 1961.

Yamaguchi, Z. Acta Orientalia Hungarica. 34. 1980.

ANCIENT CULTURE

Tucci, G. Tibetan Painted Scrolls. 1949.

Stein, R.A. Tibetan Civilization. 1972.

Haarh, E. The Yar-lung Dynasty. 1969.

A-ZHA, ZHANG-ZHUNG, SUM-PA, MI-NYAG

Hoffman, H.H.R. Zeitschrift der Deutschen Morgenländischen Gesellschaft. 117. 1967.

Hoffman, H.H.R. Central Asian Journal. 13. 1969.

Lalou, M. Acta Orientalia Hungarica. 15. 1962.

Pelliot, P. T'oung Pao. 20. 1920.

Stein, R.A. Bull. de l'École française de l'extrême orient. 44. 1951.

Stein, R.A. Bull. de l'École française de l'extrême orient. 58. 1971.

Uray, G. In Proc. of the Csoma de Körös Mem. Symp. 1978.

Yamaguchi, Z. Acta Asiatica. 19. 1970.

CHRONOLOGY OF TIBETAN KINGS

Haarh, E. Acta Orientalia. 25. 1972.

Kwantwen, L. Rocznik Orientalistyczny. 39. 1978.

Petech, L. Central Asian Journal. 24. 1980.

Richardson, H.E. Bulletin of Tibetology. 2. 1965.

Tucci, G. India Antiqua. 1947.

Tucci, G. Oriens Extremus. 1962.

Wylie, T. Central Asian Journal. 8. 1963.

ERA OF THE DHARMA KINGS

Bacot, J. Mélanges chinoises et bouddhiques. 3. 1934–35.

Bacot, J. Introduction à l'histoire du Tibet. 1962.

Hoffman, H.H.R. Tibet. A Handbook. 1973.

Uray, G. Acta Orientalia Hungarica. 26. 1972.

Voegl, C. Nachrichten der Akademie der Wissenschaften, Göttingen. I. Philologische-historische Klasse. Nr. I. 1981.

THE TIBETAN EMPIRE

Beckwith, C. In Tibetan Studies in Honor of Hugh Richardson. Michael Aris and Aung San Suu Kyi, eds. 1980.

Demiéville, P. Le Concile de Lhasa. Bibliothèque de l'Institut des hautes études chinoises. 7. 1952.

Herrmann, A. An Historical Atlas of China. 1966.

Jackson, D. Kailash. 6. 1978.

Uray, G. Acta Orientalia Hungarica. 10. 1960; 15. 1962; 21. 1968.

Uray, G. In Prolegomena to the Sources on the History of Pre-Islamic Central Asia. 1979.

Uray, G. in Tibetan Studies in Honor of Hugh Richardson. M. Aris and Aung San Suu Kyi, eds. 1980.

FOREIGN RELATIONS

Basham, A.L. The Wonder That Was India. 1954.

Beckwith, C. Central Asian Journal. 21. 1977.

Beckwith, C. Journal of the American Oriental Society. 99. 1979.

Hoffmann, H.H.R. Asiatische Studien. 25. 1971.

Ligeti, L. in Études tibétaines dédiées à Marcelle Lalou. 1971.

Majumdar, R.C. ed. History and Culture of the Indian People. 1951–69.

Petech, L. Study on the Chronicles of Ladakh. 1939. East and West. 12. 1961. Kingdom of Ladakh. 1977.

Richardson, H.E. Tibet and Its History. 1962.

Rossabi, N. ed. China Among Equals. The Middle Kingdom and Its Neighbors, 10th–14th Centuries. 1983.

Sperling, E. Tibet Society Bulletin. 10. 1976.

Stein, R.A. Journal Asiatique. 261. 1981.

Uray, G. In Contributions on Tibetan Language, History, and Culture. Wiener Studien zur Tibetologie und Buddhismuskunde. 10. 1983.

CENTRAL ASIA

Grousset, R. Empire of the Steppes. Eng. translation. 1970.

Klimburg-Salter, D. ed. The Silk Route and the Diamond Path. 1982.

Kwanten, L. Imperial Nomads. A History of Central Asia. 1979.

Stein, M.A. Serindia. Three volumes. 1921.

SCIENTIFIC STUDIES

ARCHEOLOGY

Academia Sinica. Institute of Vertebrate Paleontology. Atlas of Primitive Man in China. 1980.

Beijing Review. "Early Man Traces Found in Qinghai." Volume 28. no. 29. July 22. 1985.

Beijing Review. "New Discoveries Reveal Ancient Life." Volume 28. no. 20. May 20, 1985.

Chang, K.C. Early Chinese Civilization: Anthropological Perspectives. Harvard Yen-ching Institute. 1976.

Cheng Te-k'un. Archeology in China. 1959.

Jia Lanpo. Early Man in China. 1980.

Monumenta Archeologica. Volume Six. Chinese Archeological Abstracts. U.C.L.A. 1978.

Wang Furen and Suo Wenging. Highlights of Tibetan History. 1984.

ENVIRONMENT

Academia Sinica. Hsi-Tsang Chung-pu Ti Chih-pei. June, 1966. Translation: Vegetation of Central Tibet. Joint Publications Research Service. U.S. Department of Commerce. 1967.

Academy of Sciences, U.S.S.R. The Physical Geography of China. Volume Two. "The Tsinghai-Tibetan Region." Praeger Special Studies in International Economics and Development. 1969.

Central Intelligence Agency. People's Republic of China Atlas. Directorate of Intelligence. Office of Basic and Geographic Intelligence. 1977.

Climap Project. "The Surface of the Ice Age Earth." Science. 191 no. 4232. 1976.

GEOGRAPHY

Defense Mapping Agency Aerospace Center. St. Louis Air Force Station, Missouri. Operational Navigational Charts. 1972–1980, 1:1,000,000. G7, G8, G9, H9, H10, H11.

Survey of India and Army Map Service. 1941–1961. Asia 1:1,000,000. NH44, NH45, NH46, NH47, NH48, NI44, NI45, NI46, NI47, NJ47, NJ48.

World Atlas. Rand McNally. 1981.

GEOLOGY

Academia Sinica. Geological and Ecological Studies of Qinghai-Xizang Plateau. Proceedings of Symposium on Qinghai-Xizang (Tibet) Plateau, 1980. Two volumes. 1981.

Academy of Sciences, China. The Roof of the World. Ed. Zhang Mingtao. 1982.

Alpine-Himalaya Region. Selected Papers of the International Geodynamics Conference held in Katmandu, Nepal, 1978. Special Issue. Tectonophysics. 62. 1980.

Beijing Review."Geologists Discover Secrets of Tibet." Volume 29. no. 34. August 1986.

Chang Cheng-fa, et al. "The Geological History, Tectonic Zonation and Origin of Uplifting of the Himalayas." Institute of Geology. Academia Sinica. 1977.

Gansser, A. Tectonophysics. 62. 1980.

Mitchell, A.H.G. Journal of the Geological Society. 138, II. 1981.

Proceedings of the Lectures by the Seismological Delegation of the People's Republic of China. Jet Propulsion Laboratory. California Institute of Technology. Prepared for NASA. 1976.

Conference on Tibet. Geology. 5 no. 8. August, 1977.

Wang, Chi-yuen and Yao-lin Shi. Nature. 298 no. 5. 1982.

GENERAL REFERENCE

ANCIENT GEOGRAPHY AND PLATE TECTONICS

Dietz, Robert S. and John C. Holden. Journal of Geophysical Research. 75. 1970.

Dott, R. H. and R. L. Batten. Evolution of the Earth. 1981.

Scotese, Christopher R., *et. al.* Journal of Geology. 87. 1979.

Seyfert, C.K. and L.A. Sirkin. Earth History and Plate Tectonics. 1979.

Ziegler, A.M., *et. al.* Tectonophysics. 40. 1977.

COSMOLOGY

Barrow, John D. and Joseph Silk. The Left Hand of Creation, The Origin and Evolution of the Expanding Universe. 1983.

Weinberg, Steven. The First Three Minutes, A Modern View of the Origin of the Universe. 1977.

EVOLUTION

Ciochon, Russell L. and Robert S. Corruccini. New Interpretations of Ape and Human Ancestry. 1983.

Lewin, Roger. Human Evolution. 1984.

Stanley, Steven M. The New Evolutionary Timetable. 1981.

WORLD HISTORY

Barraclough, Geoffrey, ed. Times Atlas of World History. 1983.

Calder, Nigel. Timescale. An Atlas of the Fourth Dimension. 1983.

McNeill, William H. A World History. 1979.

GLOSSARY

SCIENTIFIC TERMS

DATING METHODS The chief method to estimate the ages of rocks and archeological remains is radiometric dating. Most rocks contain radioactive atoms of chemical elements such as carbon, potassium, or rubidium. From their known rates of natural radioactive decay (half-life), the age of the sample can be calculated.

GEOLOGICAL TIMESCALE Dates for the earliest history of the earth are provided by radiometric dating. Periods after 570 million years ago when fossils become abundant can be defined by stratigraphic sequences of fossils. The bed in which a fossil is found can then be radiometrically dated to develop a timescale. These dates are still being adjusted as new evidence is evaluated, and completely standard dates have not been agreed upon by researchers around the world.

GONDWANALAND A cluster of ancient land masses that formed the southern part of Pangaea. See Pangaea.

MEGALITHS Large stones set into the ground in various patterns in ancient times. These arrangements may have had religious or astronomical purposes.

MICROLITHIC TOOLS Small blades chipped from stone, used to make harpoons, spears, or other complex tools. Often associated with Middle Stone Age cultures that arose at the close of the Ice Age.

ICE AGE A number of ice ages have occurred in the history of the earth, the most recent one beginning 3.25 million years ago and drawing to a close about 12,000 B.C., though the present warmer period appears to be only an interglacial. Causes of this ice age seem related to variations in the earth's orbit, precession, and tilt. Climatic changes may also be related to reversals in the magnetic poles, variations in sea level, changes in the atmosphere, cosmic impacts, and the arrangement of continents.

MOUNTAIN BUILDING Intensive mountain building appears to take place at intervals in geologic history. These active periods are thought to correspond to the assembly of supercontinents as continental plates join together. Periods of active building are interspersed with lulls that appear to represent the completion of supercontinents.

PANGAEA A supercontinent that formed between 350 and 220 million years ago. The initial configuration was complete by 260 million years ago, and was followed by a westward slide that brought Africa close to North America by 220 million years ago. This is the "classic" configuration of Pangaea. The breakup of the supercontinent began by 210 million years ago as the Atlantic Ocean began to open, separating the Americas from Africa and Europe.

PREHISTORIC PEOPLES Anthropologists have identified several types of prehistoric man. *Homo habilis* (2 million years ago), *Homo erectus* (1.6 million years ago to 200,000 years ago), *Homo neanderthalensis* (120,000 to 40,000 years ago), and modern *Homo sapiens* (40,000 years ago to present). A number of fossils intermediate between *Homo erectus* and modern man have been called archaic *Homo sapiens.*

STONE AGE Human culture opens with the Old Stone Age, the Paleolithic, beginning 2.5 million years ago and lasting until 10,000 years ago. The Paleolithic is divided into three eras. Lower Paleolithic (early Old Stone Age): 2.5 million to 200,000 years ago; Middle Paleolithic (middle Old Stone Age): 100,000–40,000 years ago; Upper Paleolithic (late Old Stone Age): 40,000–10,000 years ago. The Paleolithic is followed by the Mesolithic (Middle Stone Age): 10,000 years ago until farming practices began, marking the opening of the Neolithic (New Stone Age).

UPLIFT OF THE PLATEAU Though the Tibetan plateau had formed by 45 million years ago when India collided with Asia, rapid uplift did not begin until 2–3 million years ago. Experts offer different heights and rates of uplift. For example, according to different sources, the height of the Himalayas 3 million years ago was: 2000 meters; 3000 meters; 4000 meters; or even 5000 meters.

HISTORICAL TERMS

'A-ZHA A culture or kingdom connected with the bSe tribe according to Tibetan records. 'A-zha was conquered by Tibet in the seventh century. Modern scholars connect 'A-zha with the T'u-yü-hun people, who established a kingdom in the northeastern corner of the plateau in the fourth century.

A-MI MU-ZI-KHRI-DO The ancestor of the Tibetan tribes, according to the Po-ti-bse-ru. His three different wives gave rise to the four great tribes of Tibet.

AṀŚUVARMAN King of Nepal (r. 576–621), whose daughter Khri-btsun married Srong-btsan-sgam-po.

BON A religious tradition that was practiced in Tibet before the introduction of Buddhism. Bon lineages continue to the present time.

BRU-SHA The Gilgit region along the Indus river in modern-day Pakistan. It was part of the Tibetan Empire in the eighth century.

CH'ANG-AN Capital of T'ang China. Invaded by Tibet in 763 A.D.

CH'IANG Tribes said by Han dynasty records to have been living west of China as early as 1500 B.C. T'ang dynasty accounts suggest they might be ancestors of the Tibetans. Ch'iang tribes still live in the eastern borderlands.

DATING SYSTEM The Tibetan dating system used since the eleventh century is based on cycles of sixty years. Within one cycle, each year is indicated by an animal and an element. Twelve animals are combined with five elements to give sixty unique year names, such as fire-dog or earth-horse.

'DUS-SRONG The thirty-sixth king of Tibet, great-grandson of Srong-btsan-sgam-po, and grandfather of Khri-srong-lde-btsan.

FOUR CULTURES See 'A-zha, Zhang-zhung, Sum-pa, and Mi-nyag.

GLANG-DAR-MA The forty-second king of Tibet, great-grandson of Khri-srong-lde-btsan. His persecution of Buddhists temporarily halted royal support for the Dharma.

GRI-GUM-BTSAN-PO The eighth king of Tibet. Having cut the cord connecting the kings to the sky, he was the first Tibetan king to die upon the earth. The first tombs date to his era.

GUNG-SRONG The son of Srong-btsan-sgam-po. He ruled Tibet for five years sometime before his father died.

HAN DYNASTY A dynasty ruling China between 202 B.C. and 220 A.D. Han records offer interesting information about the peoples around the eastern edges of the plateau.

LHA-THO-THO-RI The twenty-eighth king of Tibet, born 374 A.D. The first king to have contact with the Buddhist teachings.

HSI-HSIA See Mi-nyag.

'JANG See Nan-chao.

JO-BO-CHEN-PO Famous statue of the Buddha at age twelve, brought into Tibet by 'Un-shing Kong-jo.

KHOTAN An ancient Central Asian city-state, founded by the son of the Indian king Aśoka in the third century B.C. Known as Li-yul in Tibetan records. Li-yul sometimes referred to the whole Central Asian Silk Route region. Records on Li-yul history contain valuable references to Tibet.

KHRI-BTSUN The Nepalese queen of Srong-btsan-sgam-po, daughter of Aṁśuvarman, king of Nepal. Also known as Bal-mo-bza' or Bal-bza'.

KHRI-SRONG-LDE-BTSAN The thirty-eighth king of Tibet. He established the Buddhist teachings as the religion of the land.

KIM-SHENG KONG-JO The Chinese queen of Mes-ag-tshoms, daughter of the prince of Yong, who was the brother of Emperor Chung-tsung (r. 705–710 A.D.).

LINEAGE OF KINGS Tibetan kings are commonly traced back to gNya'-khri-btsan-po, the first of the "Seven Khri." They were followed by the "Two sTengs," the "Six Legs," the "Eight lDe," and the "Five bTsan." lHa-tho-tho-ri was the last of the "bTsan." This line split after Glang-dar-ma. Forty-two generations are commonly counted, though different reckonings give forty or forty-one.

LITTLE KINGDOMS A group of twelve or more small kingdoms, rgyal-phran, that were in existence before the time of the first king, gNya'-khri-btsan-po, who united them. Lists of little kingdoms are given in many old documents.

MANG-SRONG The thirty-fifth king of Tibet, grandson of Srong-btsan-sgam-po, father of 'Dus-srong.

MAṆI BKA'-'BUM Text by Srong-btsan-sgam-po.

MES-AG-TSHOMS The thirty-seventh king of Tibet, the father of Khri-srong-lde-btsan. Also known as Khri-lde-gtsug-btsan.

MI-NYAG A culture or kingdom connected with the lDong tribe, according to Tibetan records. One Mi-nyag exists in the south around lCags-la and Dar-rtse-mdo; another existed in the north beyond Koko Nor in the eleventh century, and was known to the Chinese as Hsi-hsia.

MONKEY DESCENDANTS According to many Tibetan histories, the Tibetan people are descended from an unusual monkey and a rock demoness, a brag-srin-mo.

MU-NE-BTSAN-PO The thirty-ninth king of Tibet, son of Khri-srong-lde-btsan. His reign was very brief.

GNAM-RI-SRONG-BTSAN The thirty-second king of Tibet, father of Srong-btsan-sgam-po. Also known as Slon-mtshan.

NAN-CHAO A kingdom southeast of Tibet in modern-day Yunnan. Ally of Tibet until 789 A.D. Known in Tibetan records as 'Jang or lJang.

NONHUMANS Ancient rulers known as mi-ma-yin, including gods, demons, nāgas, and other types of beings. They lived long before the time of the first king.

GNYA'-KHRI-BTSAN-PO The first king of Tibet. 247 B.C. seems a likely date for his reign, though sources differ. His lineage, which endured until the ninth century, originated outside Tibet, according to most sources—in India or in the heaven realms.

OḌḌIYĀNA A region identified by most modern scholars with the Swat valley in modern-day Pakistan.

'OD-SRUNG Son of Glang-dar-ma. He ruled portions of central Tibet after the fall of the Tibetan Empire in the mid-ninth century.

PADMASAMBHAVA The Oḍḍiyāna Guru invited to Tibet by Khri-srong-lde-btsan in the mid-eighth century.

PĀLA DYNASTY Ruling dynasty in Bengal and northern India in the late eighth and ninth centuries.

DPAL-'KHOR Son of 'Od-srung, who lived in the beginning of the tenth century. His son Khri-lde Nyi-ma-mgon left central Tibet and established kingdoms in western Tibet.

SPU-DE-GUNG-RGYAL The ninth king of Tibet, son of Gri-gum, who was exiled and returned to Yar-lung.

RAL-PA-CAN The forty-first king of Tibet, grandson of Khri-srong-lde-btsan. Also known as Khri-gtsug-lde-btsan.

SAD-NA-LEGS The fortieth king of Tibet, son of Khri-srong-lde-btsan, who reigned after his brother Mu-ne died. Also known as lDe-srong and Khri-lde-srong-btsan.

BSAM-YAS The first Buddhist monastery established in Tibet. Founded by Khri-srong-lde-btsan, Śāntarakṣita, and Padmasambhava.

ŚĀNTARAKṢITA The Abbot of Vikramaśīla Buddhist university in India. He was invited to Tibet by Khri-srong-lde-btsan.

SILK ROUTE Caravan route across the Central Asian desert that opened in 112 B.C., linking east and west. Cities established in the oases along this route, such as Khotan, Kucha, Karashahr, Kashgar, became prosperous, cosmopolitan centers.

SRONG-BTSAN-SGAM-PO The thirty-third king of Tibet. He laid the foundation for the Buddhist tradition in Tibet and began the expansion of the Tibetan Empire.

SUM-PA A culture or kingdom connected with the sTong tribe, according to Tibetan records. Modern scholars suggest that Sum-pa is connected with the Su-p'i tribes mentioned in old Chinese records.

T'AI-TSUNG The second T'ang emperor of China (r. 627–649). Famous as the virtual founder of the T'ang dynasty. His daughter 'Un-shing Kong-jo married Srong-btsan-sgam-po.

T'ANG DYNASTY The dynasty ruling China (618–907 A.D.) during the era of the Dharma Kings of Tibet. Annals kept by T'ang historians offer valuable information about Tibetan history of this period.

TA-ZIG The region of Persia. Also spelled sTag-gzigs.

THON-MI SAṀBHOṬA Minister of Srong-btsan-sgam-po. He was commissioned to devise an alphabet for writing the Tibetan language.

TIBETAN TRIBES The original Tibetan tribes included lDong Mi-nyag, bSe 'A-zha, sTong Sum-pa, and dMu Zhang-zhung. Zla and dBas, as well as the lha-rigs rGo, are sometimes counted as separate tribes. The leaders of each tribe were A-spo lDong, bSe-khyung sBra, sTong A-lcags

'Bru, and dMu-tsha rKa. Many variations in spelling can be found in different sources.

TUN-HUANG A city in Central Asia (about 500 miles northwest of Koko Nor) that came under Tibetan rule in the eighth and ninth centuries. Records in Tibetan found there in 1907 are a valuable historical source for ancient Tibetan history.

TUN-HUANG ANNALS' DATES The dates in the Tun-huang annals are recorded using only the animal symbol. But the exact date can be determined by comparison with other accounts. Well-known events, such as the invasion of the T'ang capital or the arrival of Kim-sheng Kong-jo, can be matched up, thus establishing that the year-by-year Tun-huang chronology opens with the dog year. The rest of the dates follow consecutively. See Dating System.

'UN-SHING KONG-JO The Chinese queen of Srong-btsan-sgam-po, daughter of T'ai-tsung, second T'ang emperor.

UIGHURS Turkish tribes northeast of Tibet who allied with T'ang China against Tibetan forces in the eighth and ninth centuries.

WESTERN TURKS Tribes controlling much of western Central Asia by the seventh century, though they lost ground to T'ang dynasty expansion. Allies of Tibet from the time of King Mang-srong. Included various tribes such as Qarluq, Nu-shih-pi, Tu-lu, and Türgish.

YUM-BRTAN Heir of Glang-dar-ma. He ruled part of central Tibet after the fall of the Tibetan Empire in the mid-ninth century.

ZHANG-ZHUNG A culture or kingdom associated with the dMu tribe, according to Tibetan records. Also often connected with Bon. Zhang-zhung was located in western Tibet with its capital at Khyung-lung near Ti-se. It was conquered by Srong-btsan-sgam-po.

INDEX